ENDANGERED SPECIES

ELEPHANTS

ENDANGERED SPECIES

ELEPHANTS

PETER JACKSON

THE APPLE PRESS

A QUINTET BOOK

Published by The Apple Press
6 Blundell Street
London N7 9BH

ISBN 1–85076–238–4

This book was designed and produced by
Quintet Publishing Limited
6 Blundell Street
London N7 9BH

Creative Director: Peter Bridgewater
Art Director: Ian Hunt
Designer: Sara Nunan
Project Editor: Sally Harper
Editor: Michele Staple

Typeset in Great Britain by
Central Southern Typesetters, Eastbourne
Manufactured in Hong Kong by
Regent Publishing Services Limited
Printed in Hong Kong by
Leefung-Asco Printers Limited

Photographs © Peter Jackson, except the
following: p35 Peter Johnson/NHPA; p107 E.
Hanumantha Rao/NHPA

Dedication
For Adrienne, and for Paddy, Susan, Christina and David.

CONTENTS

FOREWORD

Dr Dhriti Lahiri-Choudhury, an eminent Indian academic as well as a leading elephant expert, wrote recently: 'It is difficult to imagine India without elephants; they are so much a part of India's myths, history and cultural heritage.' This is surely true on a world scale, and so it is not surprising that there has been a groundswell of public indignation at the massacre being perpetrated by ivory poachers in Africa. It is impossible to have anything to do with elephants without being fascinated by them. My good fortune has been to see them in the wild and in captivity in Africa and Asia. I have also been saddened and enraged by coming upon mutilated carcasses, not only in the African bush, but also in India, where individually-known tuskers have been killed, leaving only their scattered, decaying bones.

The slaughter for ivory is senseless. It is killing the goose that lays the golden eggs, because, if elephants are wiped out, ivory will no longer be produced. But equally grave is the problem of ensuring peaceful coexistence between such large, wide-ranging animals and people in our ever more crowded world. It is truly a 'jumbo problem', and I pay tribute to all those involved in trying to solve it. They include my colleagues in the Asian Elephant Specialist Group of the International Union for Conservation of Nature and Natural Resources (IUCN), led by Lyn de Alwis and Charles Santiapillai. Appropriately, both are from Sri Lanka, a country famous throughout history for its elephants. In Africa, the African Elephant and Rhino Specialist Group has played a leading role in elephant conservation.

The list of those with whom I have shared a devotion to elephants is too long to be repeated here. But no one interested in elephants can forget Iain Douglas-Hamilton. Not only has he revealed to the world the wonders of elephant life, but he has been a tireless crusader against indifference to one of the tragedies of our times, enacted before our eyes in the press and on television. The tide may now be turning, but, without political commitment and public support for conservation, the elephant could still vanish from the Earth.

Peter Jackson

INTRODUCTI

LEFT: A tusker in Virunga National Park in eastern Zaire. The tusks, pointing downwards and almost straight, are characteristic of the African forest elephant.

An Indian fable tells of six blind men who were asked to describe the elephant. One felt its flank and declared that an elephant was a wall; another its ear, and decided it was a fan; another its leg, and pronounced it to be a tree; the fourth felt its trunk, and was convinced that an elephant was a snake; the fifth decided that the tusks indicated that it was a spear; and the last felt the tail and declared: 'I know what an elephant is – it is just a rope.'

The modern view of the elephant is equally varied: to some it is a magnificent animal, symbolic of the world's wild areas. To others it is a work animal, capable of amazing feats of strength; others regard it as a source of ivory, hide and meat. Some may even view it as a rather ridiculous animal, fit only for eating buns at the zoo or performing tricks at the circus; while others may see it as a serious agricultural pest. Whatever the view, it is evident that, as humans take over elephant habitat, the interests of both animals need to be reconciled.

Elephant spend most of their lives grazing and browsing in order to maintain their enormous bodies. If confined to a small area, they rapidly destroy all the vegetation. They then starve to death. Thus they need a great deal of living space. Geneticists estimate that at least 500 inter-breeding animals are required in any one area to maintain the natural evolutionary potential of a species. Five hundred elephants would require living space covering anything

BELOW: Intricate works of art on sale in Hong Kong display the skill of Chinese ivory carvers – and the threat to the elephant population.

RIGHT: Children enjoy a ride on an African elephant in the zoo in Basel, Switzerland. But most zoo riding elephants are Asian.

BELOW: A wild tusker walks peacefully through a camp at Mana Pools, on the bank of the Zambezi River in Zimbabwe.

from 500 to 5,000 km^2 (190 to 1,900 sq. miles), depending on how much elephant food was available. Virtually no area of this size can be left solely for elephants in a world ever more crowded with people. Coexistence between humans and elephants is inescapable, but this has its problems. Agricultural crops, such as sugarcane, rice and oil palm, are extremely attractive to elephants. Herds can devastate them in a few hours, perhaps ruining a small farmer and posing serious financial problems for plantation industries. It is important, therefore, to plan and manage elephant reserves in a way that will minimize conflict between people and elephants.

Elephants inspire both affection and awe. Despite their amiable appearance, they can be dangerous animals. Possessing immense power, they are intolerant of disturbance and opposition. In India alone, 200 or more people are killed by

'Wild elephants have killed 7 people'
Elephants going wild in N Bengal forests
Elephants and electric fences
If the 350 elephants of Lansdowne are to survive the onslaughts of civilization and development, not only will every village have to move out but the army will have to be given its marching orders.
Usha Rai investigates.
Wild elephants kill 8 tribals
HT Correspondent
No solution to elephants' depredations in Dooars
Copters to be used for driving away elephants
53 Killed by wild elephants
Woman, child trampled by wild elephant
Elephants driven back

elephants every year. On the other hand, they are capable of surprising sensitivity in their movements. A lone wild male at Mana Pools, in Zimbabwe, regularly walked through a tented camp, stepping delicately over guy ropes and causing little damage. He completely ignored the campers, who were frozen with apprehension at their picnic tables. In India, local people welcomed a herd which moved recently into Andhra Pradesh State, where there had been no elephants for many years. They saw the elephants as a manifestation of the elephant-headed Hindu god, Ganesh, despite the fact that that they were aggressive and killed people. Some of the people were killed because they went too close to the elephants to worship them.

LEFT: Elephants are easily tamed for work.

BELOW: They were widely used in warfare in Asia, and by Romans, Greeks and Carthaginians. During the recent Indo-China war, the Viet Cong employed them for transport.

It is a remarkable phenomenon that such a powerful animal should be so easy to tame. Within a short time, a captured wild elephant is capable of learning commands and applying its strength and skills in the service of humans. For several thousand years it has been used for transport, war and ceremonial. For more than a century, it has been a key worker in the timber industry. But, in all these spheres, the elephant's role is declining or has already gone. Emperors, kings and maharajahs no longer exist to maintain vast elephant stables for their armies and ceremonies. Internal combustion engines power transport in almost every type of terrain. And while thousands of elephants are still employed in the forests, mainly in Burma, their tasks of moving and loading logs are being taken over by machines. Zoos and circuses can support only a limited number of elephants. They are expensive to maintain.

Probably no other wild animal matches the popularity of the elephant. A leading British economist, uninvolved in nature conservation, wrote to the chairman of the World Wide Fund for Nature, the late Sir Peter Scott: 'I have never seen them, but it makes me happy to think that there are still wild elephants in Africa and whales in the southern seas.' But, if elephants are to survive, they need the support of the people who live among them, as well as of all those who believe that, without elephants, the world would be a poorer place.

ABOVE: Caparisoned elephants are still used in Indian pageantry. The Dussehra festival in Mysore is famous for its elephant parade.

ONE

GIANTS OF THE EARTH

LEFT: Wild elephant herds are captured by being driven into a stockade. They are then individually roped and trained for work.

Fifty million years of trial and error have gone into the making of the African and Asian elephants we know today. They are the sole survivors of more than 300 species of trunked animals which have trodden the Earth. Some of them were far larger than those we now see. They roamed not only over most of Africa and Asia, but also lived in North and South America.

Fossil evidence shows that one of the elephants' ancestors was an animal, named *Moeritherium,* which resembled a large pig or a small hippopotamus. Remains of this animal were first found at Lake Moeris, near Fayyum, in Egypt. It lived during the Eocene epoch, around 50 million years ago, in swamps at the edge of the Tethys Sea. Although far removed in appearance from elephants as we know them, *Moeritherium* had features that foreshadow the evolution of the whole order of trunked and tusked animals, which we call Proboscidea. The second incisor teeth in the front of the upper and lower jaws of *Moeritherium* were enlarged and resembled tusks. It did not have the prominent canine teeth seen in carnivores. Instead, it had primitive molars, or grinding teeth. The skull of *Moeritherium* shows affinities with those of modern hyraxes, small African mammals like large guinea pigs, and with sea-cows (dugongs and manatees). Neither of these animals looks at all like modern elephants, but they are their closest living relations.

ABOVE: Surprisingly, the diminutive hyrax is a close relative of the elephant.

Modern elephants have tusks which grow only from their upper jaws. But their extinct predecessors, which have been divided into five families, evolved an extraordinary variety of tusks. Some followed *Moeritherium* in developing elongated incisors, which extended like pincers. In others, such as *Gomphotherium*, they became like scissors. The related *Platybelodon*, found in Mongolia, and the North American *Ambelodon*, grew large shovel-shaped lower jaws and tusks, which must have been used for digging. *Deinotherium*, including the species *D. giganteum*, which was nearly 4 m (13 ft) tall at the shoulder, had tusks only in the lower jaw and they curved downwards. Very powerful, they were also probably for digging.

Proboscidea that migrated to Eurasia and went on to cross a land bridge over the Bering Straits to America gave rise to the mastodons. North America has many mastodon remains. Finds of *Mastodon americanus* even include some preserved skin and long, reddish hair. Their tusks, growing from the upper jaw, resemble those of modern elephants. In South America lived *Cuvieronius*, which had spiral enamel on its tusks. Old World stegodons, which closely resembled modern elephants, grew tusks 3 m (9.8 ft) long. But the tusks of the species *Stegodon ganesa*, whose remains have been found in Siwalik hills at the foot of the Himalayas, grew so close that there was no room between them for the trunk.

The Elephantidae, the family to which today's African and Asian elephants belong, appeared about 5 million years ago. Originating as *Primelephas*

LEFT AND BELOW: Of the several species of hyrax, some are expert tree climbers; others prefer rocky habitats.

in Africa, it evolved into more than 20 species. Gomphotheres were probably the ancestors of *Primelephas*.

Mammoths evolved from *Primelephas* in Africa and spread into Eurasia and the Americas. They were closely related to living elephants, and especially to the Asian elephant. The earliest known mammoth, *Elephas trigontherii*, stood 4.5 m (14.8 ft) at the shoulder, distinctly taller than a present-day African elephant. It was the largest of all the elephants. Almost as big was the imperial mammoth, *Mammuthus imperator*, of North America. One of the best known of all the species was the woolly mammoth, which was about the same size as the average Asian elephant – 3 m (9.8 ft). Not only was it widely depicted in Stone Age cave paintings of about 20,000 years ago, but whole carcasses, complete with flesh and skin, and with food in the stomach, are still uncovered from time to time in the frozen wastes of Siberia. Known scientifically as *Mammuthus primigenius*, the woolly mammoth was covered with long, dense, dark hair. Some had tusks 5 m (16.4 ft) long, which spiralled inwards. The hair, together with a thick layer of body fat, enabled the woolly mammoth to survive in the intense cold of the Ice Ages. It had a high domed head and a long sloping back. Stone Age people were inspired to paint mammoths, but they also hunted them, and may have contributed to their extinction around 10,000 years ago. They were not the last of the non-living elephants to disappear, however, for remains of *Cuvieronius* in Ecuador have been dated at AD 200–400.

The African elephant *Loxodonta africana* is considered to have changed least from *Primelephas*. It had one 'cousin', *L. atlanticus*, which became extinct. The evolutionary line leading to the Asian elephant *Elephus maximus*, however, included many species, now extinct, which were found throughout Eurasia and America, as well as in Africa. Fossils of dwarf species, such as *Elephas falconeri*, about 1 m (3.3 ft) at the shoulder, have been found on Mediterranean islands.

Although humans have been implicated in the extinction of some of these early elephant species, changes in the Earth's climate, which affected vegetation and the availability of water, were primarily responsible for their decline and disappearance.

BELOW:
The now-extinct mammoth was related to the Asian elephant. Stone Age people depicted it in cave paintings. This specimen was found in 1969 in the Swiss Jura mountains.

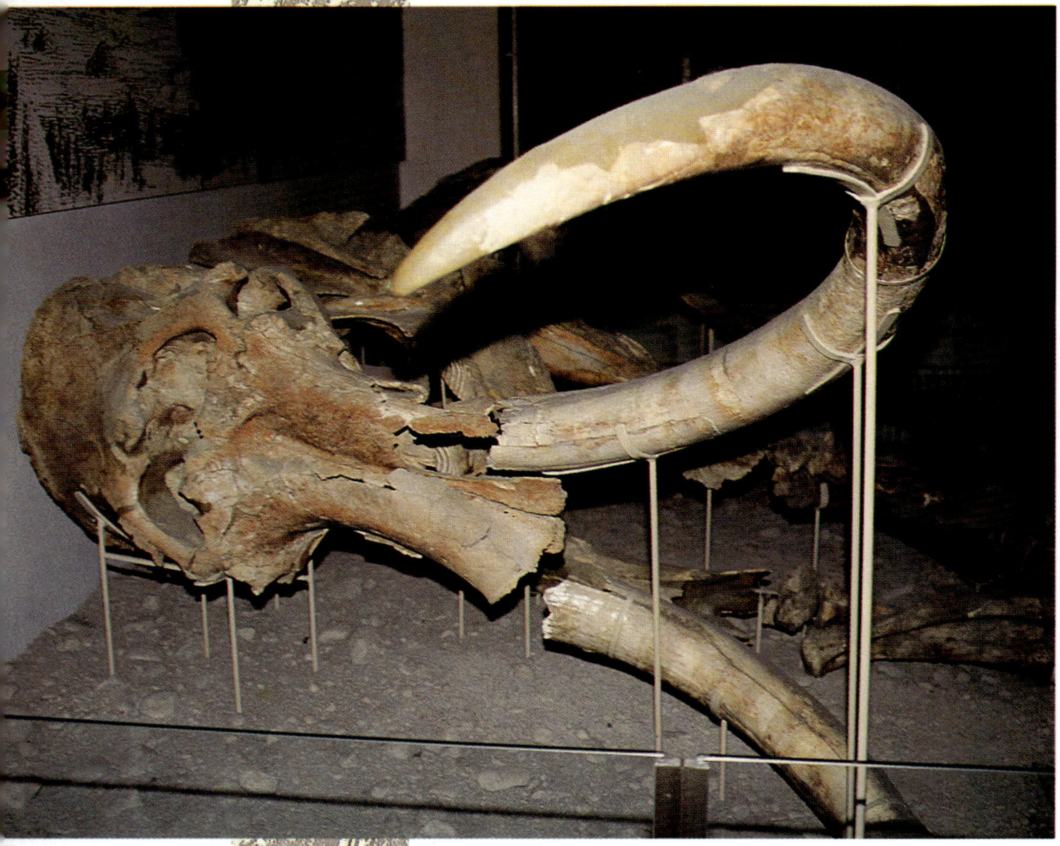

TWO

ELEPHANTS TODAY IN AFRICA AND ASIA

LEFT: The versatile trunk can reach for food, suck up water for drinking and showering, tear up grass to eat, or lift and manipulate both large and tiny objects.

RIGHT: Elephants support their heavy heads on short necks, which consist of seven vertebrae, just as those of humans. Despite their long necks, giraffes (below) also have only seven neck vertebrae.

Several animals have tusks, but only elephants have that distinctive trunk. How did it come about? Rudyard Kipling relates in his *Just-So Stories* that it resulted from a tug-of-war with a crocodile. He was clearly drawing on an ancient Hindu myth. More prosaically, it seems that the trunk evolved to enable the elephant to reach the ground! An elephant's head became massive to accommodate its tusks, and, in the African species, may weigh 1,000 kg (2,200 lb). To manoeuvre such a weight on a long neck would need impossible muscles, and so an elephant's head has remained closely attached to its body by only the shortest of necks. By contrast, the giraffe has a relatively very small head, and can

LEFT: An Asian elephant demonstrates the effectiveness of his 'portable shower'. *BELOW:* An African elephant brushes soil from the roots of grass before eating it.

support it on a long neck. Surprisingly, the giraffe's elongated neck contains exactly the same number of cervical vertebrae (seven) as the elephant's.

In the course of evolution, the elephant's nose and upper lip combined to form a mobile appendage to gather grass, leaves and fruit and convey them to its mouth. The trunk is primarily a nose for breathing, which can also be raised to test the air for scent when danger is suspected. It acts as a pneumatic hose capable of sucking and blowing liquids when drinking and showering, and to collect dust to blow over the body. No other mammal can use its nose in this way. The trunk can be used as a trumpet and sounding board. As a highly flexible working tool, it is almost as versatile as the human hand.

Elephants can tear up grass with their trunks, bang the dirt off the roots, and

RIGHT: An elephant uses its trunk to greet and caress its companions.

BELOW: A young African elephant folds its trunk back for protection when charging.

convey it to their mouths. They can select fruits, break off branches, and hoist huge logs. With equal facility, they can locate and pick up a small coin and place it with extreme accuracy. Elephants even dig in sandy river beds with their trunks in search of water. They use their trunks as a delicately expressive organ of affection to greet and caress fellow elephants. Trunks can also function as weapons, although in a serious charge the elephant will coil its trunk out of the way. Despite the importance of the trunk to elephants, they have been known to survive even when it has been cut short.

Elephant tusks are simply enlarged incisor teeth in the upper jaw. They are composed of dentine with a cap of enamel that wears off. A baby elephant has milk tusks, which fall out when it

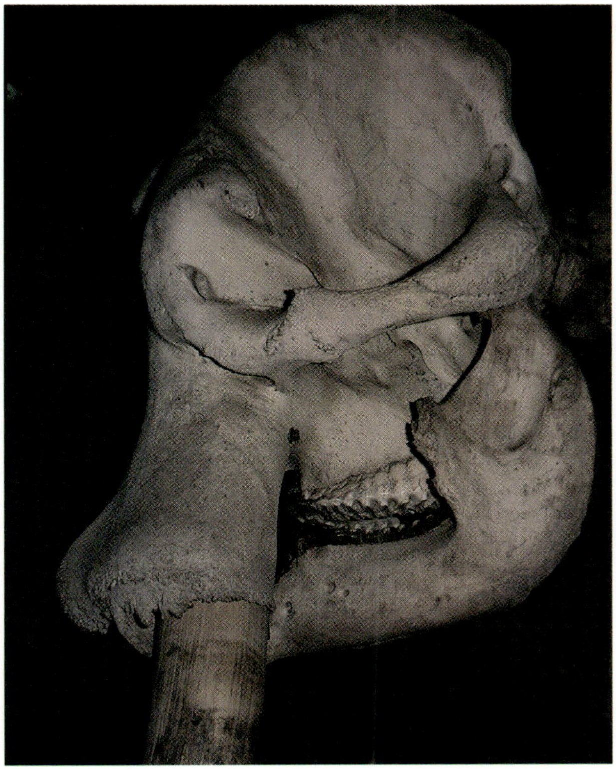

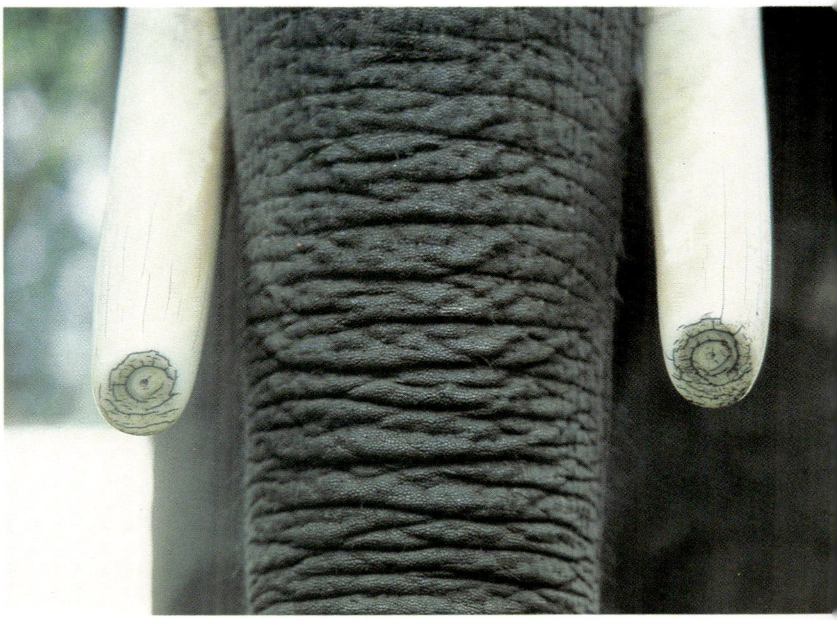

is one year old and the tusks are about 5 cm (2 in) long. A quarter of the length of an elephant tusk lies within the socket, where it is held by a mass of tough, fibrous tissue. Pulp, composed of vascular tissue and nerves, fills the cavity within the tusks to just beyond the lip. A thin nerve canal, known as the 'heart', runs through the core right to the tip. Tusks continue to grow throughout an elephant's life, so that big tusks denote relatively old elephants. Female tusks are lighter than those of males.

Although they can be, and are, used as weapons, tusks are primarily tools. Earlier elephant forms, including mammoths, used tusks to sweep away snow when feeding, and the tusks of some were adapted for digging and scooping. Modern elephants use their tusks to

TOP LEFT: Elephant tusks are enlarged incisor teeth deeply rooted in the upper jaw.
ABOVE: The dental nerve runs through a tiny channel in the centre of the tusk.
LEFT: Tusks are used to scrape bark from trees for food, but the trees may die.

RIGHT: In Sri Lanka, the few male elephants which have tusks are often captured for display and ceremonial. Males without tusks (below) may be even more powerful than tuskers.

prise bark from tree trunks and to dig. Just as people are right- or left-handed, elephants use one or other tusk by preference, usually the right. Tusks suffer from wear which, in old elephants, may exceed growth. They can get broken.

Both sexes of the African elephant bear tusks, but in the Asian elephant only the male has them. The females have small 'tushes', which seldom protrude beyond the lip. Not even all male Asian elephants grow tusks. The percentage failing to do so varies in different regions. In Sri Lanka, only 5–7% of males have tusks. Just across the water, in southern India, 90% have them, while, north of the Ganges River, only 50% of the males are tuskers. Absence of tusks does not diminish an elephant's status in the pecking order. Tuskless males, known in India as *makhnas*, are often bigger and more aggressive than their tusked fellows. Tuskless males sometimes appear among African elephants. The presence of several in a herd or local population indicates a genetic cause.

The longest tusks recorded for a modern elephant came from one shot in the eastern Congo (now Zaire), which were obtained by the New York Zoological Society in 1907. The right tusk measured 3.5 m (11.4 ft) on the outside curve, and the left 3.35 m (11 ft). Their combined weight was 133 kg (293 lb). However, an elephant shot below Kilimanjaro in 1897 had tusks weighing together 211 kg (465 lb), although they were shorter, measuring 3.11 m (10.2 ft) and 3.19 m (10.5 ft).

Perhaps the best-known tusker of all time was Ahmed, because he lived in

LEFT AND BELOW: Because of his massive tusks, Ahmed was proclaimed a 'national treasure' in Kenya. He died a natural death in 1974. His skeleton, tusks and hide have been preserved for display at the Nairobi Museum.

the age of television and worldwide communications. Ahmed was renowned for his splendid tusks, and was even declared the largest living elephant. The late President Jomo Kenyatta of Kenya declared Ahmed a 'national treasure' and special guards protected him from poachers in the Marsabit Mountain Reserve. He died of old age in 1974. He proved not to have been a giant, measuring only a modest 3 m (9.85 ft) at the shoulder. His tusks, too, magnificent as they were, weighed only 134 kg (296 lb), well below the estimates made during his life of 150 kg (330 lb).

The remainder of an elephant's teeth are also remarkable. Elephants are completely vegetarian. Like cows, horses and sheep, they grind up vegetation so that it can be digested. Elephants grind in a forward and backward

ABOVE: Elephant teeth consist of a series of transversal plates to grind food. *RIGHT:* Elephant skulls are lightened by a honeycomb structure of air passages.

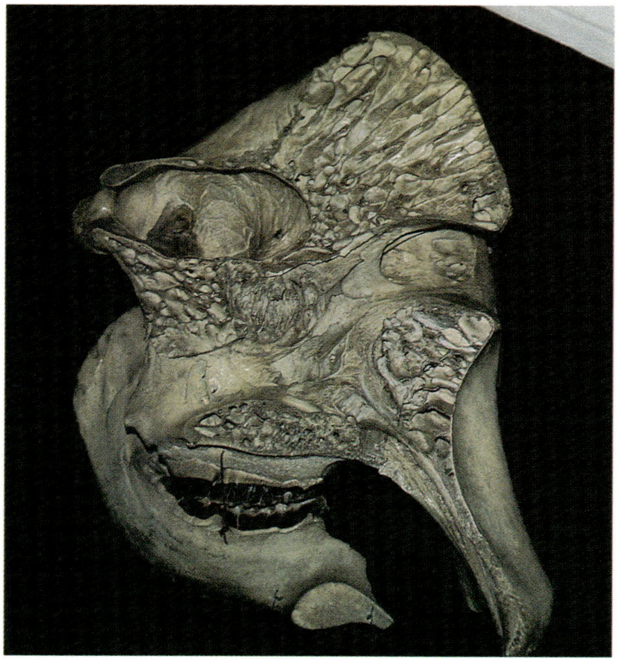

movement of the jaws, and fossils show the evolution of more efficient surfaces to the molars. They consist of lateral plates of hard enamel and dentine, bound together by cement, which are highly effective grinding surfaces. Prehistoric *Moeritheriums*' teeth had only crude double humps. But, as time went on, the teeth of various elephant species became longer, with more and thinner cross plates. The plates, or laminae, number up to 14 in the African elephant, and are lozenge-shaped. In Asian elephants the edges of the plates are parallel.

Unlike humans and other animals, elephants have a 'queue' of six molars in each jaw, which move forward as

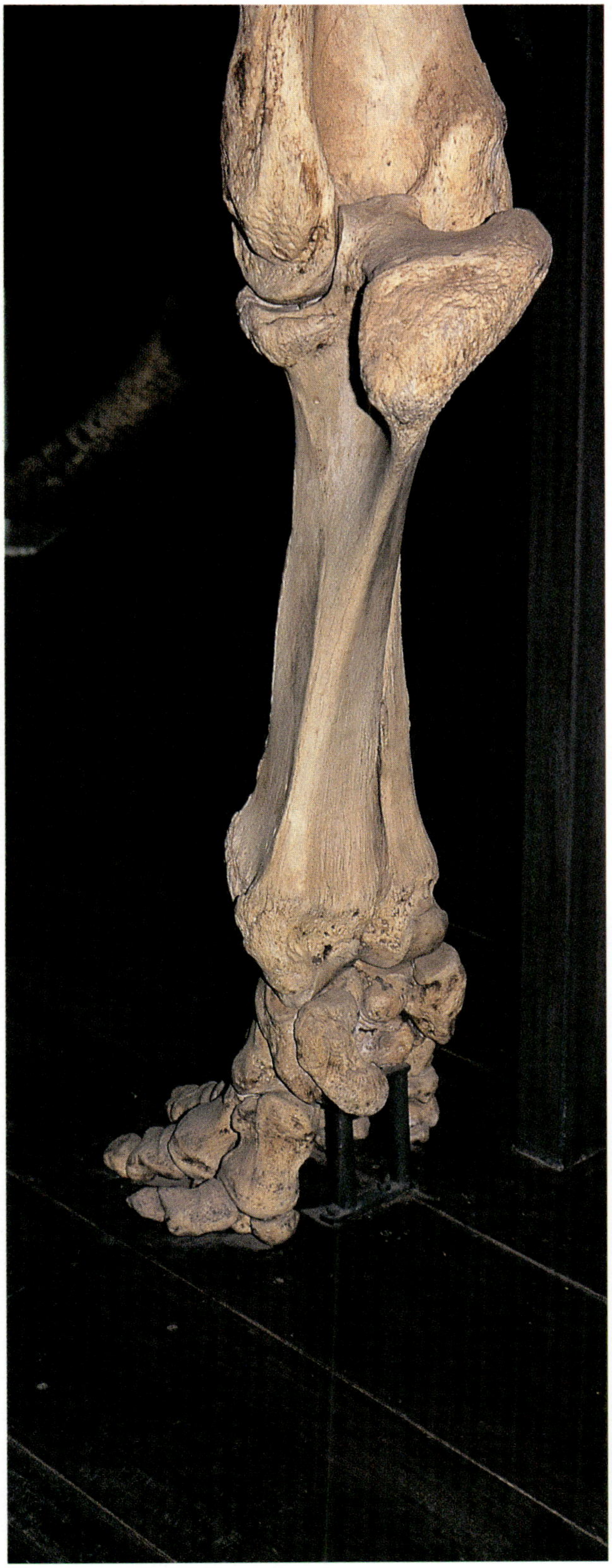

each successive set wears out and breaks off. Only one set is fully in use at any one time. Occasionally, elephants have a seventh molar, but normally, with the final wearing of the sixth molar in their sixties, they are doomed because they can no longer feed.

Elephant skulls are massive, but much lighter than might be expected. A large part of the forehead is honeycombed by air passages, which lighten the bone. The skulls of African elephants are more massive than those of Asian elephants because they support heavier tusks. The brain is located at the back of the skull and is larger than that of any other mammal, although smaller than the human brain in relation to the elephant's size.

To support the great weight of the body, elephants evolved pillar-like legs. The limb bones are strengthened by a filling of spongy bone, which replaces the marrow found in other bones. The

LEFT AND BELOW: Elephants have evolved massive leg bones to support their great weight. They walk on the balls of their feet, with their heels supported by wedges of fibrous tissue.

big mass of the foot consists of a circular pad with the toes showing as nails at the front – usually five in the forefeet and four in the hind. A wedge of fibrous, fatty tissue supports the rear of the foot and spreads the weight over the ground. It has been calculated that the stiletto heel of a woman's shoe carries as much weight in proportion to surface area as an elephant's foot. Elephants leave very little trace of their passing on hard soil. Given their bulk, they cannot gallop, but their fast walk can move them at about 25 kph (15.5 mph). The circumference of the forefoot is roughly half the shoulder height of an elephant. This enables trackers to gauge the size of their quarry by measuring the footprints.

Both African and Asian elephants have wrinkly skins, which hold water and keep the skin moist, so that the body is cooled by evaporation. Here again, the species differ. The African elephant's skin has deeper and finer channels than the open honeycomb-like pattern of the Asian elephant's, and thus can retain more moisture. This is an advantage in the far drier conditions in which African bush elephants live. The wrinkled skin also holds the wet mud, with which elephants spray themselves, much more efficiently than a flat surface.

Smell is the most highly developed sense of elephants. Elephants pick up surrounding scents on the breeze. If the scent disturbs them, they raise their

BELOW: Elephants can cause serious problems for people who settle in their formerly untouched forests.

RIGHT:
After bathing, elephants often spray themselves with mud to kill skin pests and keep cool.

trunks and point the tips in that direction. They constantly feel and smell each other with their trunks. In this way males pick up the pheromones which identify females ready to mate.

An elephant's eyes are about the same size as those of human beings. Long eyelashes shield the eyes and help to produce the benign look so often noticed. Like reptiles and some other mammals, elephants have a protective, nictitating membrane (third eyelid) which can cover the eyeball. Vision is limited in scope: while poor in bright light, it is acute in the forest shadows.

The ears of the African bush elephant are particularly impressive. In anger, they are spread to magnify the animal's already great size and terrify intruders. But they also serve as regulators of body heat. A network of blood vessels on the back of the ear permits the cooling of hot blood coursing from the arteries to

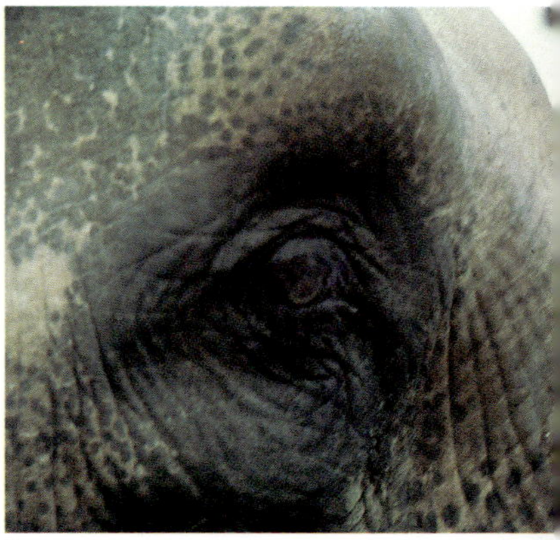

LEFT: An African elephant lifts its trunk to scent the air.
ABOVE: Elephants have long eyelashes.

ABOVE: Sheltered by the protective bulk of its mother, a baby African elephant sniffs the scent of an intruder.

the veins. Scientists have found a difference in temperature of up to 19°C between the arterial blood coming from the heart and the returning venous blood. The elephant may stand with its back to the wind, so that it blows over the blood vessels and takes away the heat. If there is no wind, it will flap its ears like fans, thus generating a current of air. At night, or during cold or wet weather, the elephant holds its ears close to its body to retain warmth.

The African forest elephant and the Asian elephant (also a forest animal) have smaller ears than the bush elephant, probably because they are usually protected from the sun by the trees. But they still use their ears to some extent for cooling.

The large ears might suggest that an elephant has acute hearing. However, it does not appear to be especially highly developed. This has made all the more intriguing the apparently coordinated movements of elephants separated by considerable distances. In May 1984, Dr Katherine Payne was watching some Asian zoo elephants in Portland, Oregon, when she felt a 'palpable throbbing like distant thunder', which reminded her of the throbbing produced by pipe organs. She wondered whether the elephants were producing similar vibrations without the sound being audible to the human ear. Ultrasensitive equipment revealed that this was indeed true. Furthermore, the skin on the elephant's forehead could be seen

RIGHT:
Angry elephants spread their great ears to magnify their terrifying size when charging.

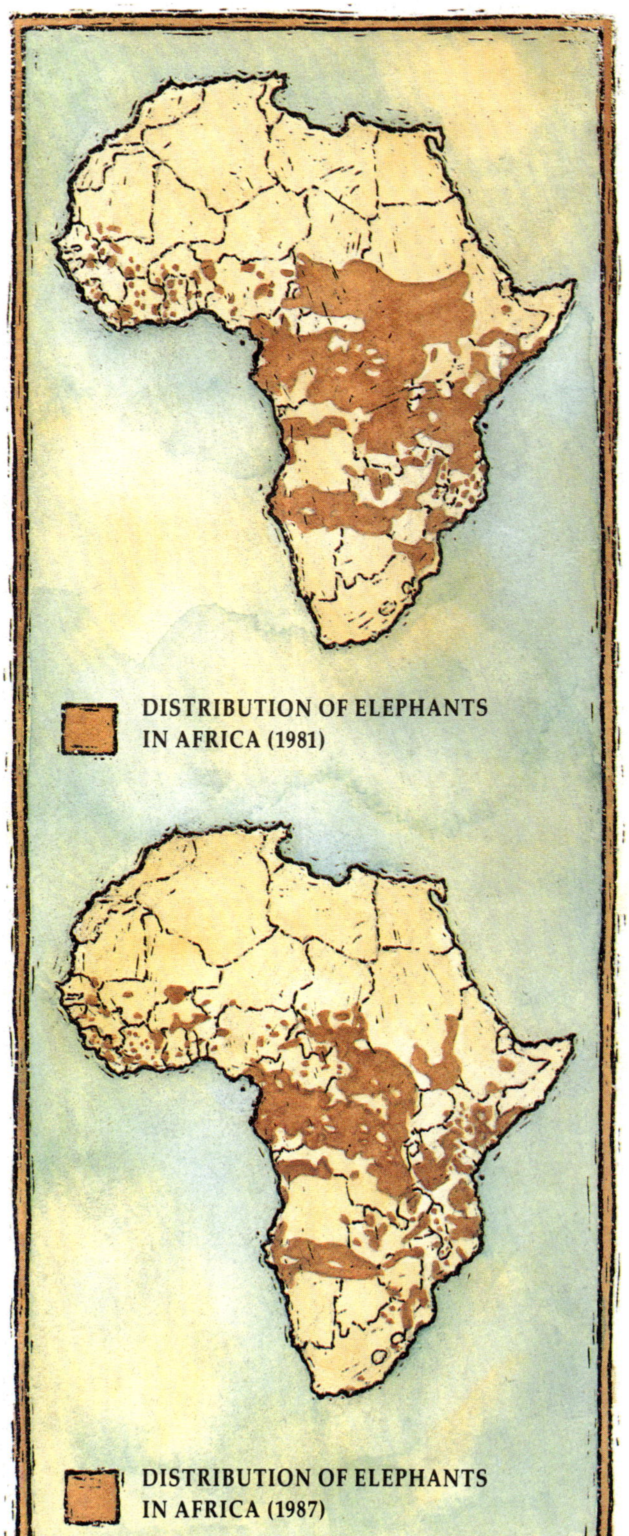
DISTRIBUTION OF ELEPHANTS
IN AFRICA (1981)
DISTRIBUTION OF ELEPHANTS
IN AFRICA (1987)

PROBABLE DISTRIBUTION OF THE AFRICAN ELEPHANT AROUND 1600

vibrating with the sound. Zoo recordings over a period of 17 days produced clustered calls, marking events such as the arrival and departure of keepers and the relocations of elephants in their display and resting areas. Although the observers again noted throbbing in the air, many calls were only found when the tapes were speeded up to produce audible frequencies.

Tests at Amboseli, in Kenya, and Etosha, in Namibia, revealed that the subsonic frequencies of 14–35 hertz (cycles per second) enabled elephants to communicate and coordinate movements over at least 4 km (2.5 miles). Females 'talked' to their babies and to other members of the herd. When female mating calls were played back, male elephants stopped in their tracks and headed in their direction.

Apart from subsonic communication, elephants make a range of about 10 audible sounds. The best known is trumpeting, but they also snort, bark, cry, growl and rumble. High-frequency sounds, such as trumpeting, are usually a sign of stress, or excitement.

Unlike those of most mammals, the testicles of a male elephant do not descend into a scrotum, but are deep in the body close to the spine. The huge penis, too, is withdrawn within the body, leaving a slit between the legs so like the female vulva that it can be hard to determine an elephant's sex in the wild. When mating, the penis erects into an S-shape which hooks into the female vagina.

The digestive system of elephants is simple and resembles that of horses. They do not ruminate (chew the cud); the stomach stores food, which then passes through long intestines – 19 m (62.3 ft) in a large African bull – where the cellulose of the coarse vegetation is broken down and the food digested. The faeces are rounded like small footballs, and include a great deal of undigested matter. Seeds are distributed in the faeces and sprout new vegetation.

African elephants are of two distinct subspecies. The bush, or savanna, elephant, *Loxodonta africana africana,* is familiar because it is easily visible on the grassy plains of East and Central Africa. It is a mainstay of the wildlife tour industry and a staple of television films about Africa. Its close cousin, the forest elephant, *Loxodonta africana cyclotis,* is little known. Living in the dense rain forest of the Zaire basin and parts of West Africa, it is seldom seen, and few studies of its life have been carried out.

The forest elephant has an average shoulder height of less than 3 m (9.8 ft), compared with that of the bush elephant, which may be as tall as 4 m (13.1 ft) and weigh 4,700 kg (10,360 lb). The forest elephant's ears are also smaller. Their rounded shape gives rise to the scientific name *cyclotis.* Tusks are thinner and point downwards. In transition areas between bush and forest, elephants show a mix of the characteristics of both subspecies.

Since elephants are associated with lush vegetation and permanent water, it is a surprise to find some in arid country. A few are found in the Sahel, the southern fringe of the Sahara, where they trek over a thousand kilometres to find water. They may be relics of the elephants which inhabited north Africa 2,000 years ago. In southwestern

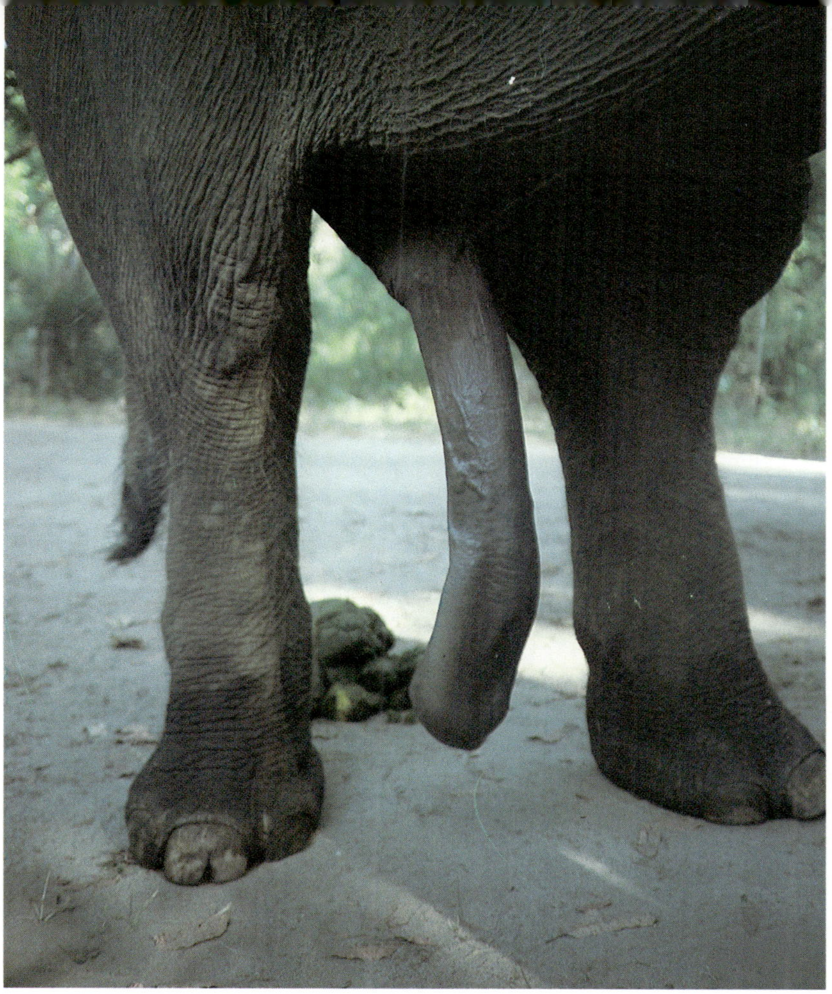

LEFT: The elephant's penis withdraws completely within the body, but when erect it is more than one metre long and S-shaped to reach the female's vagina.

BELOW: Balls of elephant dung contain much undigested matter.

ABOVE: Young male African elephants climb the bank after swimming across the Zambezi.

Africa, 'desert' elephants are found in Namibia. They dig with their tusks and trunks in sand rivers to find water flowing under the surface, and manage to find sufficient vegetation to support their large bodies.

An even more bizarre place to find elephants is underground. But at Mount Elgon, on the Kenya-Uganda border, herds march in file into the darkness of Kitum Cave, using their trunks to feel their way among the rocks, like blind people with sticks. Investigating the phenomenon, biologist Ian Redmond found that they were seeking mineral salts. Saltlicks are common places to find wild animals, although they get some salts in their plant food. However, in the area of Mount Elgon, the salts are leached from the vegetation by the heavy rainfall, but there are deposits in the mountainside. Redmond found that the elephants chipped the mineral-rich rock from the walls of the cave and chewed it. He speculated that they had thereby 'excavated' the cave over, perhaps, 100,000 years. Not surprisingly, the Kitum elephants have well-worn tusks.

The Asian elephant is smaller than its African cousin, but it still weighs over 4,000 kg (8,800 lb) and may reach more than 3.5 m (11.5 ft) in height. It is easily distinguished because its ears are much smaller and its back rounded, unlike the saddle-back of the African species. The Asian elephant has a single 'finger' on the upper tip of the trunk, compared with two in the African species, and twin bumps on the forehead instead of a single dome. The Asian elephant is the one most commonly seen in zoos and circuses. The difference between subspecies in Sri Lanka, India and continental South-east Asia, and Sumatra are negligible.

THREE

FAMILY LIFE AMONG ELEPHANTS

LEFT: An elephant's eyes shine in torchlight.

RIGHT: Crocodiles share the same habitat as elephants, but will generally avoid them.
BELOW: Giraffes have to spread their forelegs wide to drink.

All is calm around the waterhole in the late African afternoon. A few gazelle graze nearby. A crocodile basks on the muddy shore, its mouth open. A plover picks morsels from between the crocodile's teeth – it looks daring, but the bird is confident that the crocodile appreciates the service and will not snap it up. Waders probe the mud for worms. Two giraffes emerge from the forest, where they have been nibbling the tops of the acacia trees. With stately tread and haughty looks they head for the water, spread their forelegs wide and lower their heads to drink. The raucous cry of a fish eagle rings out.

The surrounding plain appears deserted, but gradually a dark mass in the distance dissolves into a herd of 50 elephants marching purposefully towards the waterhole. Antelopes scatter from their path, sped by the waving threat of huge heads and trunks. Approaching the water, the herd quickens its pace, the young scurrying ahead with excitement. The giraffes recoil and move discreetly away. The crocodile slips below the surface of the pool. The elephants spread out in the shallows, trunks extended to suck in vast quantities of water, which they squirt down their throats. Babies duck under the water, only their trunks visible. Thirsts quenched, the elephants shower their bodies. Some plunge deeper into the water, creating great waves as they roll around.

From the nearby forest, another herd emerges. They pause, trunks raised, tips bent forward to sniff the scent. Elephants in the water also lift their trunks and sniff towards the newcomers. Old friends and relatives are

LEFT: The raucous cry of the fish eagle resounds throughout Africa.

RIGHT:
A herd of elephants approaches a waterhole in Zimbabwe for the usual evening drink and bathe.

ABOVE AND RIGHT: Elephants drink frequently, and appear ecstatically happy when bathing.

recognized. The second herd moves to the water, and lines up alongside the first arrivals. Greetings are exchanged with waving trunks, sometimes intertwined in affectionate caresses.

A third herd appears. As it moves to the water, the first group turns as one, and withdraws a short distance. There it almost vanishes as the elephants complete their toilet by throwing dust over their bodies. An air of contentment envelops the scene.

A large bull elephant arrives alone, drinks and bathes, and then wanders among the herds, his trunk exploring the females as he seeks a mate.

As night descends, all the elephants withdraw to the forest and spread out. The sound of cracking branches tells that they are feeding.

Three-quarters or more of an elephant's day is spent feeding. It occurs in three main bouts – one in the morning, another in the afternoon, and a third around midnight. In between, time is spent digesting, resting, drinking and bathing. During the wet season they eat mainly grass, and turn to browse in the dry weather. Elephants often enter swamps and lakes to eat the vegetation.

Normally, elephants drink once a day, but they can go for longer periods without water. Some elephants, trapped accidentally in a corral, managed to last 14 days without drinking. It left them very emaciated, but they survived. Apart from drinking, elephants obviously enjoy bathing and showering themselves with their trunks. They plunge under the water, and roll around voluptuously. They swim strongly, using their trunks like 'snorkels' to

LEFT: When herds meet at the waterhole, they greet each other with trunk caresses. ***BELOW:*** Elephants swim across a reservoir in India using their trunks as 'snorkels'.

breathe. In January 1982, a boatman sighted three elephants swimming far from shore in Lake Kariba on the Zimbabwe-Zambia border. Wildlife guards followed them, estimating that they moved at about 2 kph (1.25 mph). In all, they swam 35–40 km (22–25 miles) and spent at least 27 hours in the water.

Elephants sleep for relatively brief periods; at night, during the early hours of the morning and during the heat of the day. They may lie down, mostly at night, but they also sleep standing, their trunks resting on the ground.

Elephants are endearing in their close social organization and their obvious affection for each other. In the past, it was assumed that the herd consisted of a bull accompanied by his harem and their offspring. However, some hunters and others familiar with elephants in the wild suggested that the herds were led by females. It was only 25 years ago that a young Scotsman unravelled the truth about elephant life. In the mid-1960s, Iain Douglas-Hamilton went to live with the elephants in Tanzania's Lake Manyara National Park. Day after day he followed them, observed them feeding, resting and drinking, and learned to recognize individuals by the shape of their tusks and the ragged edges of their ears. He soon established that herds were family groups, led by an older female. Her companions were sisters, daughters, nieces and their offspring. Males in the herds were young. As they matured they left to live alone or with a few temporary companions.

Douglas-Hamilton's studies were followed by other scientists. Some have

ABOVE: African elephants enjoy a midday nap in the shade. Some lie down, but most stand with drooping heads, resting their trunks on the ground.
RIGHT: A baby Asian elephant affectionately rubs against its mother.

ABOVE: Male elephants live apart from herds of females and young. They often form loose companionable groups; these are not permanent.

now spent much of their lives with the elephants of a particular area, and got to know each individual. They have watched matings, births and deaths, and thus been able to build up family histories. Family groups are stable. They may split into sub-units, specially during the dry season, when food is scarcer and more scattered, but the family tie remains and all rejoin the herd from time to time. A nucleus group usually numbers fewer than 10, but when the whole family gets together there may be 40–50. Even larger gatherings of elephant clans, numbering a hundred or more, occur.

The matriarch is absolutely dominant, and obedience comes naturally to her herd. She is usually recognizable as the largest in the herd, because elephants grow throughout their lives and she will be the oldest. Now in her fifties or sixties, the matriarch is the repository of the herd's wisdom. From long

experience, she knows when and where the trees are fruiting, or the grass is lush; where there is water to drink and to bathe in. She leads the herd along traditional routes that have been trodden by many elephant generations. The herd follows her obediently. At a waterhole, they enter the pool together, and leave together. Stragglers are unusual. She is fiercely protective, and is in the forefront if any threat emerges to the herd or an individual member.

Scientists have found that the matriarch will usually not let them near to a junior member of the herd that they have tranquillized. Other herd members will back her defiance. But if the matriarch herself is tranquillized, the herd loses its coherence and fails to come to her defence.

At the death of the matriarch, her role as leader is taken on by another

ABOVE: The matriarch leading the herd signals that the time has come to leave the water, but a youngster snatches a last drink.

LEFT: During courtship the male lays his trunk along the female's back as they prepare to mate.

experienced female. This may mean that her particular family merges with a sister group.

When the males mature they leave the female herd to form loose companiable groups, which are not permanent. A bull may wander away from a group to join another or remain solitary for a long time.

The female oestrous cycle takes 2–3 weeks, and the fertile period lasts less than 48 hours. However, elephant twins are occasionally seen. Bulls detect the crucial period for mating by sniffing the female genitals with their trunks. They then put their trunks in their mouths. Like many other male animals, they possess a Jacobson's organ in the roof of the mouth, which enables them to detect the presence of female hormones.

No lengthy courtship has been observed. The female may move away from the male at first, but soon they entwine trunks. From the rear the male lays his trunk along the female's back and then stands and places his forefeet on either side of her spine. His S-shaped penis finds and penetrates her vagina. Copulation lasts less than one minute. The male does not make pelvic thrusts, which would be difficult because of his size and position. Instead, the penis itself is activated by special muscles.

Several bulls may attend and mate with an oestrous female. Fights occur until one bull becomes dominant. He is probably successful in impregnating the female, because he has sole access at the critical time of ovulation.

Gestation takes about 22 months. This proves Aristotle right when he wrote in the fifth century BC that it took nearly two years, and Pliny the Elder wrong when, six centuries later, he suggested six months. Because of her bulk, it is seldom obvious that an elephant is pregnant. Elephant owners have been taken by surprise by births.

Barbara and Walter Leuthold were able to observe and photograph a birth sequence in the Samburu Game Reserve in Kenya. One cow was seen to be walking backwards and bending her hind legs. Suddenly, the herd burst into squeals, rumbles and trumpeting, and a calf, still wrapped in the foetal membrane, lay on the ground. The mother at once started to remove the membrane with her trunk, tusks and forefeet, while the rest of the herd milled around. Only the matriarch came close. With the membrane off, the mother nudged and dragged the calf, which stood briefly after 20 minutes. An hour after the birth, the mother had eaten the membrane and the calf was well on its feet and able to move off with the herd. At this stage, and for the first year of life, it is able to walk under its mother's belly.

Elsewhere, the expectant mother has been seen to separate from the herd, but to be accompanied by another female, who has been called the 'auntie'. The auntie helps tend the newborn. All the herd show their pleasure at the new arrival, surrounding it and feeling its hairy little body with their trunks.

The female elepant's teats are located between the forelegs, equivalent to the position of those of human females, rather than around the abdomen as in most mammals. Suckling – the baby uses its mouth, not its trunk – continues for two or three years. At

ABOVE:
A baby Asian elephant supports itself on a forefoot to reach his mother's breast.

first, the young elephant has to raise itself with a foot against its mother's leg to reach the teat. Growth is fast. From a birth weight of about 120 kg (265 lb) an African calf grows to 1,000 kg (2,200 lb) by the time it is six years old.

Elephants are firm but tender mothers. The mother frequently strokes her calf with her trunk, and uses it as an aid in difficulty. The trunk is also used to chastise, baby elephants being as exuberant and naughty as human children. The calf gets the care and attention, not only of its mother, but of all the females in the herd, including older sisters and cousins. All

TOP: Baby elephants are often hairy. *LEFT:* A herd of elephants bunches together when confronted by a vehicle in northern India.

will make sure that it stays with the herd and is protected from danger. If a threat occurs, the herd bunches, with the young secure in the middle. Small calves retreat beneath their mothers. For baby elephants, the main danger is from lions, which can throw a herd of elephants into near panic. For adults, only humans present a lethal threat. In the past, elephants were hunted with bow and arrow and spear. Today, the poacher is often armed with an automatic rifle, capable of killing the whole family in seconds.

Most of the information about elephants in the wild has been obtained by studies of the African bush elephant,

ABOVE: Elephants are seldom vulnerable to predators because of their size, although lions in Africa and tigers in Asia may kill babies.

LEFT: Humans are the greatest threat, especially those with automatic weapons, such as this one held by a park guard.

ABOVE: Loss of habitat is a major threat to elephants. Fires may destroy large areas, which take many years to regenerate.

which is relatively easily visible in its savanna and open woodland habitat. Far less is known in detail about the African forest elephant, which is found in the dense equatorial forests, and the Asian elephant, which also lives in thick forests. However, observations of the forest species indicate that the general pattern of life is the same.

Weaning begins when the baby is a few months old. As with most wild animals, many baby elephants die in their early years. Mortality depends on many factors, but the availability of food is obviously critical. Droughts can reduce food supplies, and elephant herds themselves can create famines by destroying vegetation in crowded conditions. Floods, fires and other disasters take their toll, as do accidents and disease. Kenya's Tsavo National Park became a centre of controversy in the 1960s over whether elephants should be culled because habitat was being ravaged by drought and direct destruction. A study by Richard Laws suggested that mortality in the first year was as high as 36%, and 10% from one to five years. On the other hand, Douglas-Hamilton's research in Lake Manyara National Park indicated that first year mortality was only 10%, dropping to an average of 3–4% annually thereafter.

Local conditions also affect the age of sexual maturity, which can vary from the tenth year of life to 17 or 18 years of age. Stress is an important factor. The age of maturity is higher when elephants are in crowded conditions, as happens nowadays when they are harassed by poachers and increasingly forced into reserves by the spread of human settlement. These conditions also result in a lower birth rate, with longer periods than the average four years between pregnancies. Again food availability is critical. This was confirmed by detailed research carried out on the reproductive system of wild elephants by John Hanks in the Luangwa Valley in Zambia. He found that most females did not ovulate during the dry season, when there is less food.

Asian bull elephants suffer from a condition known as *musth*. The symptoms include a discharge from a gland between the eye and the ear. The elephant becomes irritable and violent. Working bulls have to be chained for several weeks until the condition passes. Wild bulls in *musth* appear to be more successful in mating. After years of debate, and following studies by Americans Joyce Poole and Cynthia Moss, it has now been accepted that

African bull elephants also experience *musth*, during which they provoke fights, usually over access to females, with other males not in *musth*.

In parallel with human females, elephants experience a menopause in their forties, although one Asian female gave birth at 61, just after she had been retired from timber work. It is also in the forties that the final set of the six succeeding molar teeth comes into use. When they are worn out, the elephant can no longer masticate its food and is doomed. Rarely, an elephant may have a seventh set of molars. This was discovered during culling operations in Uganda, when four jaws out of 385 had the extra molar. Hanks did not find the seventh molar in any of the Luangwa

ABOVE: Elephants can be their own enemies when they destroy trees by tearing off the bark. ***LEFT:*** Bull elephants become violent in a condition called 'musth'. This one has been tranquillized so that it can be chained up.

ABOVE: Badly injured in a fight, a dying elephant takes to water.

Valley elephants he was studying. But his wife, Carol, did. They were using an elephant jaw as a seat in their camp latrine. One day she turned it over – and there was a seventh molar, the only one in 1,236 specimens.

When death is imminent, do elephants go to traditional graveyards? In legend they do, but there is no evidence that this actually happens. Elephants often go to waterholes and rivers when they are dying, and their bodies disintegrate there. Floods wash the bones into basins and hollows, which then give the impression of graveyards. Herds of females and immatures may die in groups near the last available water during droughts. There are also places where deadly gases seep from soils. Only recently many people died in the Congo when gases erupted from a lake. Elephant herds too may perish when this happens, again creating the impression of a graveyard.

AN AFRO-ASIAN ELEPHANT

African and Asian elephants have been classified in different genera, *Loxodonta* and *Elephas*. There are obvious differences in their form, including their skeletons. Animals in different genera are not expected to be capable of producing offspring. But, in 1978, to the surprise of her keepers, an Asian elephant at Chester Zoo in England gave birth to a baby, whose father was African. The keepers were not even aware that the pair had mated nearly two years earlier. The calf, premature, but apparently quite healthy, lived for 11 days before dying of a gastric problem that sometimes affects young animals. It had a remarkable mixture of African and Asian elephant features. The forehead had both the single dome of the African species, and the twin domes of the Asian. The body shape was mainly African, but had a hump in the centre of the back, like an Asian elephant, and a hump at the rear like an African. It is the only record of an Afro-Asian elephant.

FOUR

ELEPHANTS AND PEOPLE

LEFT: An Asian elephant ridden by mahouts at sunrise.

No other animal has had such a close relationship with people as the elephant, and still remained wild. Although never domesticated in the same way as the horse, it has been tamed and used as a beast of burden for thousands of years. Carved seals from the Indus Valley civilization, which thrived in the Indian subcontinent 4,000 years ago, show elephants with cloths flung over their backs, indicating they were tamed. Tribal people in India and other parts of Asia had probably subjugated elephants even earlier. Knowing of the ease with which elephants can be controlled, it is possible to imagine stray calves being adopted and then put to work. The Vedas, ancient Hindu writings from between BC 1500 and 1200, already mention tamed elephants. At first, they were called 'wild beast *(mriqa)* with a hand *(hastin)*', later, just *hastin*. To this day, elephants are called '*hathi*' or '*hasti*' in India and Sri Lanka.

RIGHT: India's pride in its elephants is reflected on a postage stamp.
FAR RIGHT: An Etruscan plate depicts a war elephant of Pyrrhus, ruler of ancient Epirus. The Romans panicked his elephants by sending squealing pigs among them.

Wild elephants were also found over large areas of ancient China. Some were kept in zoos or tamed and used for riding and transport. In Africa, 5,000-year-old rock drawings in the Nile Valley depict elephants being hunted, but not tamed. Later, the Egyptian pharaohs were keen elephant hunters, mainly for ivory, in the Nile Valley and Somalia. They also hunted the Asian elephant, then widespread in Syria. The tomb of Thutmose III, who campaigned in Mesopotamia in the fifteenth century BC, has a drawing of a small elephant, clearly Asian, being led by a rope. A stele of 879 BC records that King Assurnasirpal of Assyria captured as many as 30 elephants at a time in a pit. He also kept a herd of elephants in a zoo. Possibly through such extensive hunting and capture, as well as spreading human settlement, the elephant appears to have become extinct in Syria by 500 BC.

The Bible contains a detailed description of the elephant and its habitat. 'Behold now Behemoth' says the Book of Job, 'he eateth grass like an ox . . . he lieth under the shady trees, in the covert of the reed and ferns . . . his strength is in his loins . . . his bones are as strong pieces of brass . . . his bones are like bars of iron.' The final verse, 'his nose pierceth through snares', clearly refers to the trunk, for trunk snares are known from history and are used today.

In Greek literature, Herodotus was the first to mention elephants in his account of his journey up the Nile Valley in the fifth century BC. Apparently he never saw the animal, but he writes that there were elephants in

Libya, to the west, and Ethiopia, to the east. Two centuries later, Aristotle was able to give a detailed (and remarkably accurate) description of the elephant in his *History of Animals*. He had been tutor to the young Alexander the Great, who took naturalists with him when he launched his expedition to northwest India in 331 BC, and they may have briefed Aristotle. But the greater significance of Alexander's Indian adventure was that it led to the introduction of elephants as a major arm in Mediterranean wars.

Alexander's first encounter with war elephants was in a confrontation with the Persian king, Darius, who deployed 15 elephants among his cavalry and scythed chariots. They fell into Alexander's hands. As he marched eastward, he captured or acquired more. By the time he reached India and fought his great battle with Porus, ruler of present-day Punjab, he was able to field 136 elephants. Porus had an even bigger contingent – 200 to support his 60,000 cavalry and infantry. Nevertheless, Porus was defeated and abandoned 80 elephants to Alexander. One account tells that Porus was defended by his elephant when he slipped from it to the ground, wounded and exhausted. Alexander captured Porus, but later restored him to his throne.

Alexander's use of war elephants spread among contending rulers around the Mediterranean. Pyrrhus, king of Epirus in Greece, used them extensively in his campaigns against the Romans. In one battle, the Romans panicked Pyrrhus's elephants by sending squealing pigs among them. Similar tactics were to be used frequently in battles against elephants, which caused havoc to their own side when they ran amok.

The Egyptians employed both African and Asian elephants, in war and in ceremonies. In an epic battle at Raphia, Syria, on 22 June, 217 BC, Ptolemy IV fielded 73 African elephants against the Seleucid king, Antiochus II, with 102 Asian elephants. Whereas King Porus a century earlier had ridden his elephant bare-backed, the animals were now equipped with turret-like howdahs, from which soldiers lunged at each other with pikes. Polybius, the Greek historian, wrote that most of Ptolemy's elephants were afraid, 'as is the habit of African elephants, for, unable to stand the smell and the trumpeting of the Indian elephants, and terrified, I suppose, also by their great size and strength, they immediately run away from them before they get near them.' The African elephants ran over their own men, and Antiochus gained the day. Since Ptolemy's elephants were awed by the Asian species, they must have been the forest subspecies, smaller than the African bush elephant, and then still found in north Africa.

In the western Mediterranean, Romans and Carthaginians used elephants in their wars in north Africa, Italy and Spain. In 218 BC, a year before the battle of Raphia, Hannibal led a corps of 37 elephants from Spain across the Rhone and over the Alps to outflank the Romans. He fought his way successfully through snow and hostile tribes to reach the northern Italian plains with his elephant corps intact. But when the Romans fought back, Hannibal lost most of his elephants,

ABOVE: Elephants once roamed over large areas of China. This statue decorates the Imperial Palace in Beijing.

FAR LEFT: Elephants are among the large stone sculptures lining the Imperial Way to the Ming tombs, near Beijing.

largely because of winter cold and rain. Elephant reinforcements, however, bolstered his campaign. Coins issued by Italian towns to celebrate his victories mostly show elephants with the rounded ears and saddle-back of the African species, but some coins clearly show the small ears and humped back of the Asian elephant. It is thought that Hannibal may himself have ridden one.

Romans later imported elephants for public spectacle rather than war. They were paraded in triumphal or religious processions, or tested for their prowess by fighting bulls in the arena. In Pompeii's lavish Games in 55 BC, 18 elephants fought a mock battle against African war prisoners, armed with spears, until the crowd wept with compassion for the animals. Later spectacles preferred to show elephants' skill in performing tricks similar to those used in circuses today.

European crusaders probably encountered elephants during their campaigns in Palestine to capture the holy places for Christendom, since they founded an 'Order of the Elephant.' The order was revived in AD 1693 by King Christian V of Denmark, who issued a presentation coin showing an Asian elephant bearing a castle howdah filled with warriors. But the elephant's warrior days were already over in Europe; instead it had become a popular royal gift.

LEFT: Near Bhopal, in central India, a tribal artist has decorated a cave wall with elephants. *BELOW LEFT:* A brass elephant with ceremonial howdah.

An elephant, African judging from drawings, was presented to Henry III of England by Louis IX of France in 1254. It was kept in the Tower of London. In the sixteenth century, Queen Elizabeth also received an elephant from France. Willian IV had one which became a founder member of the collection at Regent's Park, the origin of the London Zoo. Covent Garden Theatre in London staged an elephant in 1810. Later it became violent and was shot by a detachment of foot guards.

King Emmanuel of Portugal presented an elephant named Hanno to Pope Leo X in 1514 when seeking the pope's ratification of his country's conquests in Africa and the Far East. Hanno seems to have shown a sense of humour, or contempt, by spraying water over the church dignitaries. Despite this *lèse-majesté*, the pope called on Raphael to design a monument on its death in 1516.

The elephant made a very good impression on Europeans. A *New Dictionary of Natural History*, published in 1785, is full of praise. It declares: 'All historians are agreed that, next to man, the elephant is the most sagacious of all creatures.' But it remarked that the elephant's physical aspect was one of 'extreme stupidity.' Among many early

BELOW RIGHT: Realistic, life-sized elephants have been carved on rocks at Mahabalipuram, in southern India.

European drawings and paintings of elephants, there is a notable one by Rembrandt in about 1637.

In Africa, people were surrounded by elephants. Inevitably, they played a role in folklore. Clans had elephant totems, and would not dream of killing their treasured symbol. Among the Shonas, the largest community in Zimbabwe, *nzou,* the elephant, is a traditional clan totem, admired for its size, strength and wonderful tusks. The tusks have themselves given rise to derivative totem names. *Ndoro,* meaning an ivory necklace, is used as a symbol of authority, and *nyandoro* is the name given to the bearer of the symbol of authority. Some Shona expressions draw on nature, as *mashambanzou* for 'dawn', which is literally 'when the elephant washes.' Elephants usually play a benign role in African folk tales.

BELOW: Elephants are revered in Thailand. The statue represents the legendary three-headed Erawan.

Throughout the elephant range in Asia, rulers amassed large elephant stables for use in peace and war. When Timur, king of Samarkand, attacked Delhi in 1398, his men were nervous at the sight of the defenders' elephants. But Timur protected his camp with ditches and pointed stakes stuck in the ground. Then he sent camels and buffaloes with blazing grass on their backs among the opposing elephant phalanx. They broke up in panic, trampling and disorganizing the Indian forces, who fled. Timur marched into the city and sacked it. According to one account, the captured elephants fell before Timur with a great moan, as if suing for quarter. Knowing the power of command that mahouts (elephant-keepers) have over their elephants, it is easy to guess who instigated the gesture.

The Moghul emperors of India in the sixteenth and seventeenth centuries kept huge elephant stables. Jehangir is reported to have had 12,000 war elephants, and a total of 40,000. He was outdone by Khmer rulers in Cambodia, who are said to have had 200,000. Jehangir's elephants were used for hunts, during which they would form decreasing circles around tigers, rhinoceroses and herds of deer and antelope, so that the riders could slaughter them. Some were employed as executioners to trample on the condemned.

LEFT AND BELOW:
Elephant-headed Ganesh is a popular Hindu deity. He brings good fortune and, as demonstrated by the scribe who wrote out the epic Mahabharata, is invoked at the beginning of books and all enterprises.

For Asian peoples, however, the elephant has had much greater significance than merely as a beast of burden or war. It has been an inseparable part of their life and culture. Ancient Hindu works refer frequently to elephants, and there is a major work on elephant lore, the *Gajasastra*. Chief of the early Vedic deities was Indra, who is depicted riding his celestial, four-tusked white elephant, Airavata. Airavata was born from the churning oceans and is revered as the ancestor of all elephants. He supports Indra as guardian of the East, one of eight elephants involved in guarding the points of the compass.

One of the most popular gods to this day is the elephant-headed Ganesh. He was the son of Siva, one of the principal Hindu deities, and Parvati. Siva is said to have cut off the boy's head in a fit of anger. Then, in remorse, he replaced it with that of the nearest animal, which happened to be an elephant. As the God of Wisdom and Remover of Obstacles, Ganesh is worshipped by Hindus at the beginning of any important undertaking. He is invoked at the beginning of

ABOVE: A phalanx of caparisoned elephants parades through Mysore, in southern India, during the Hindu Dussehra festival.

books because he is said to be the scribe who wrote down the great epic of the *Mahabharata*. His worship was spread through Southeast Asia by Hindu voyagers and settlers. Ganesh temples and images are found in Thailand and Indo-China, and on the Indonesian islands of Java and Bali.

For Buddhists, too, the elephant has special significance. Before Gautama Buddha's birth, his mother, Maya, dreamt that a white elephant entered her side. Wise men told her it was a sign that she would give birth to a great man. The white elephant, a rare form in the wild, features in many Buddhist stories and has been revered for centuries in Thailand and adjoining areas. It has even been the cause of wars. In the sixteenth century King Machachakraphat of Siam incurred the envy of King Burengnong of Burma because he had seven white elephants. The Burmese king demanded two of them

ABOVE: Keepers spend many hours decorating their elephants for ceremonial pageantry.

and the dispute led to a war in which the king of Siam was reduced to a vassal of the Burmese ruler. Five kings are said to have died in white elephant wars between rulers of Thailand, Pegu and Arakan. Even today any white elephant captured in Thailand automatically belongs to the king, who has a stable where they live in indolent luxury.

Elephants continue to be stars of oriental pageantry. In Sri Lanka, a giant caparisoned tusker, escorted by other richly decorated elephants, carries the reputed tooth of the Buddha in stately procession at the annual festival of Esala Perahera in Kandy. Many Hindu temples in southern India maintain stables of elephants for ceremonial occasions. In Mysore, southern India, the great autumn festival of Dussehra would be incomplete without its parade of elephants painted with colourful designs, and draped with rich cloth. Ceremonial elephants also carried the

British royal family and their guests at the coronation of the king of Nepal.

The Chinese passion for animal medicines included those made with parts of the elephant's body. Ivory was used as a diuretic, and for epilepsy, osteomyelitis, smallpox, jaundice and female sterility. Flesh was prescribed for bald spots; bile for halitosis; eyeball mashed in human milk for eye diseases; skin for injuries and ulcers; and bone as an antidote for poisons, as well as for vomiting, diarrhoea and poor appetite.

The most practical use of elephants today is in the timber industry. Over

ABOVE AND RIGHT: Elephants are skilled timber workers, using their great power to shift huge logs at the command of their mahouts.

5,000 trained elephants help to harvest teak in Burma's forests. Others are employed in India, Thailand and Indo-China. At the command of their mahouts, elephants drag huge logs down precipitous slopes; balance them carefully as they lift them on to trucks, and nudge them into position; push them into rivers to form rafts; and even clear logjams. During World War II, the British and Indian armies used elephants in the Burma campaign against the Japanese. Field Marshal Lord Slim wrote later: 'It was the elephants' dignity and intelligence that gained our respect. To watch an elephant building a bridge; to see the skill with which the great beast lifted the huge logs, and the accuracy with which they were placed into position, was to realize that the trained elephant was no mere transport animal, but indeed a skilled sapper (military engineer) . . . . Without them our retreat from Burma would have been even more arduous and our advance to its liberation slower and more difficult.' Even in modern warfare, the elephant was invaluable.

ABOVE:
A 19th-century drawing shows British hunters mounted on an elephant to search for wild boar.

Elephants have not only been employed in Asia. In 1914, they helped build the Long Beach board-walk on Long Island, New York.

In the past, elephants were used for hunting tigers and rhinoceroses. Nowadays, scientists find elephants ideal mounts for their studies, because wild animals are generally not disturbed by them or their riders. Elephants also carry patrols in national parks and provide transport in difficult country. For many tourists, the highlight of a visit to India or Nepal is a ride on an elephant, which can safely take them into the jungle for close-up views of wild animals.

In recent times elephants were used for transport by the Viet Cong during the Vietnam War. American helicopter gunship crews were instructed to kill any they spotted, in order to deprive the enemy of their use.

Africa has not shared the tradition of taming wild animals, which in Asia resulted in the domestication of cows, sheep, goats, pigs and chickens. Thus the African elephant is known mainly as a wild animal. However, in the former Belgian Congo (now Zaire), Belgians called in Asian specialists and their elephants to capture and train African elephants. The experiment showed that the African elephant, which had the reputation of being intractable, can be tamed. A few tame African elephants still remain at Gangala na Bodo in northeastern Zaire.

All over the world, elephants are a major attraction in zoos. Usually the

ABOVE: Scientists find elephants valuable when tracking animals in the jungle.
LEFT: In the past, elephants were safe mounts for hunters.

riding elephants are Asian, but in some places, for example, in Basel, Switzerland, children ride an African elephant. The zoo staff say they find no difference in temperament between Asian and African species.

Best known of all zoo and circus elephants is Jumbo. He came as a young calf from southern Africa to Paris, and was sent to the London Zoo in 1865, where he grew to a shoulder height of 3.5 m (11.5 ft). When he became violent, breaking his tusks at the root on one occasion, the great circusman, Phineas T. Barnum, bought him. This provoked enormous public indignation in Britain, but Jumbo was finally shipped to the USA, accompanied by his own keeper, Matthew Scott. He appeared in circuses, leading proces-

ABOVE: Guards in Manas National Park, in northeast India, patrol on elephants.
RIGHT: To see wild animals, tourists penetrate difficult country on elephant back.

sions of Asian elephants. On 15 September, 1885, he was knocked down and killed by a railway engine when returning from a circus in Ontario, Canada. His massive skeleton is in the Museum of Natural History in New York.

Most circus elephants, however, are Asian. They delight audiences by playing with their keepers, performing intricate movements in unison, standing on their heads and kicking footballs around. Their actions demonstrate the control they have over their apparently unwieldy great bodies. Standing on their hind legs, however, is not just a circus trick.

In films, trained Asian elephants are often disguised as African by attaching false ears, and even tusks. Their rounded backs soon give them away. The continuing popularity of elephants is reflected in Walt Disney's endearing baby elephant, Dumbo; his regimental line-up of comic elephants, commanded by the bumbling Colonel Hathi in the *Jungle Book;* and the lovable French cartoon character, Babar.

TOP LEFT: Even on elephants, umbrellas are useful in the rain.
LEFT: Wild elephants stand on their hind legs like circus animals to reach high branches.

ABOVE: African elephants can be tamed and ridden as easily as their Asian cousins.

FIVE

CATCHING AND TRAINING ELEPHANTS

LEFT:
Elephant for sale! Special elephant fairs are held in India and other Asian countries.

Wild elephants are captured in Asia either singly or in groups. The single target animal is usually about 10–20 years old, because at that age they are quickly tamed and have a long working life ahead. The catchers ride out on two tame elephants, called 'koonkies' in India, carrying lassoes of heavy rope. The first aim is to separate the young elephant from the herd, and especially from its mother, who will try to defend it. While one team keeps the herd and mother at bay, the other manoeuvres so that the riders can secure the noose around the young elephant's neck. As it is led away, the second elephant urges it along and holds off the mother if she tries to follow. An even more risky way of catching elephants is to creep up on foot and slip a noose around a leg. The other end is quickly tied to a tree.

Pitfalls, used mainly in southern India, may be the earliest method employed by humans for capturing elephants. Trappers dig a pit on a known elephant trail, and camouflage it with branches and leaves. An unwary elephant falls in. The trappers put a rope round its neck, and cut a ramp on one

BELOW:
A young wild elephant is captured by lassooing.

LEFT: Elephants are often captured by preparing concealed pits for them to fall into.

BELOW: A trapped elephant is roped and led up a ramp from the pit.

side of the pit, up which the elephant can climb. As it struggles up the ramp, they attach more ropes to its legs to secure it. Bull elephants are sometimes caught by enticing them to follow a tame female elephant in oestrous into a trap.

Most spectacular is the rounding up of whole herds, which are driven into a stockade. This is the 'khedda' in India, and the 'kraal' in Sri Lanka. The stockade is prepared at a strategic point in the forest, with a V-shaped funnel entrance through which the elephants can be driven. The whole is camouflaged with branches. Teams on elephants and on foot locate a likely herd and guide it towards the stockade. This may take several weeks. During the day the teams keep disturbing the herd to push it in the right direction. The aim is to get it to a place where there is water and good cover by nightfall. A line of fires stops the elephants breaking back. Once the herd nears the stockade, the final drive begins. Shouting and banging, their elephants trum-

RIGHT:
The Khedda is a spectacular way of driving a herd of wild elephants into a stockade for domestication.

peting and roaring, the catchers chase the herd through the funnel and into the stockade. A heavy wooden gate crashes into place.

Blessings are almost always invoked before the capture of elephants. In Assam, young chickens are ceremonially released at the entrance of the stockade as a token exchange of life for those that are to be taken from the wild. A stockaded herd may become quite violent, charging the fence and trying to attack the catchers gathered outside. In former times the elephants would be left without food or water, to weaken them before an attempt was made to rope and secure them. This cruel method often led to the loss of more than half the captured elephants. But, early in this century, a British forester in India, A. J. Milroy, insisted that this was unnecessary. He personally led a team of catchers on elephants into a stockade the day after capture and rounded up the wild herd successfully. Wild elephants apparently do not recognize that there are vulnerable riders on the tame elephants, and do not attack them. Losses were reduced to a mere handful. Unfortunately, the wasteful old methods still survive in Burma.

From the stockade the wild elephants are driven, a few at a time, into a small roping arena. After much jostling between the koonkies and the wild elephants as the catchers struggle to attach ropes, each elephant is led off with an escort of koonkies and tied by the legs between two trees.

At least once a day the captured elephants are escorted to water to bathe and drink. Koonkies teach them by their example to eat proferred food.

LEFT: Captured elephants are subdued by tame ones so that they can be secured by ropes. ***BELOW:*** A newly-captured elephant is led away.

RIGHT: Some elephants fight against captivity, but others (below) are quickly calmed and taught to take food by their domesticated brethren.

The catchers talk and sing around them to accustom them to human presence. Subsequently, they put each captive into a very narrow cage, which allows little movement. There they gently stroke it, while singing traditional songs telling how well it will be looked after.

Training starts with the captive firmly roped between two koonkies. One trainer stands before its head, and another at its tail. Its future mahout slips on to its back. If the elephant is violent, it gets a prod on the forehead with a sharp stick and its tail pulled. After a few days, it is calmer. The mahout sits on its neck with his toes beneath the ears. He gives the command 'advance!' and presses his feet against the back of the ears. The koonkies move forward, pulling the captive along. At the command 'halt!', they stop sharply. The captive gets a prod on the forehead, and a pull at its tail. Within a few days, it understands the command and obeys. As training progresses it learns the verbal and touch commands to turn right and left, and to move backwards. First one, and then both koonkies are withdrawn. Training continues until the tamed elephant learns a variety of orders. Eventually, most elephants learn 40 or more commands. Wild elephants do not naturally sit down. They have to be taught to do so as part of work they will do. The lesson is never forgotten. Tame elephants which have returned to the wild still sit.

BELOW:
To accustom newly captured elephants to humans, their captors stroke and sing to them for hours.

RIGHT: Trained elephants assist a captured animal to learn commands. ***BELOW:*** The mahout guides his mount by pressing his toes behind the elephant's ears.

The close relationship between mahout and elephant may last their lifetimes. The mahout's wife confidently leaves her baby and small children in the elephant's care. Woe betide an intruder. Although largely affectionate, the relationship involves firmness by the mahout to maintain dominance and control. Elephants get beaten for disobedience. Some mahouts deliberately make open sores on an elephant's head, so that a touch with a hooked goad or a stick is a sharp reminder to behave. Sometimes elephants dislike a particular person, and even kill their mahout. In one case, the mahout had left his elephant without food and drink for three days. When he returned, the angry elephant knocked him over with its trunk, trampled his body and tore it to pieces.

A particularly dangerous time is when a bull is in *musth*. He is then extremely irritable and violent. It is usual to chain a *musth* bull for the several weeks that his condition lasts. He may even get sedative foods and medicines.

A typical working day for an elephant lasts only about four or five hours. The rest of the time is spent feeding, drinking, bathing and resting. Each mahout has an assistant, whose task is to collect leafy branches, bamboo and grass for their elephant. He also prepares elephant delicacies, such as rice and vegetables wrapped in banana leaves. If necessary, medicines are hidden in the packages.

Except when working or going to water, the elephant is usually kept chained. But some are trusted to wander alone in the surrounding forest,

LEFT: Despite their power, elephants are usually calm when tethered.

ABOVE: If they particularly dislike their mahout, elephants may take violent action.

TOP: A pleasure for a captive elephant is a daily bath by his keepers.

RIGHT: Elephants make affectionate mothers.

a clanking chain attached to a foot to mark their whereabouts. Females may be mated by wild bulls when left free. The female works until she gives birth, but afterwards she has to look after her baby for some time. Later, it will accompany her when she is working. Although tame from birth, the young elephant is not capable of work for 10 years or more, and so it is a costly burden to the owner. This is a factor which weighs against captive breeding of work elephants. More captive breeding of elephants for the timber industry is likely to be encouraged in future to prevent depletion of numbers by capture of wild elephants.

ABOVE: Bathtime for mother and baby. ***LEFT:*** An elephant awaits a purchaser at the elephant fair in Sonepur, northern India.

LEFT:
Indian elephants silhouetted against the dawn light as they head for the river to bathe.

Trained elephants are sold at fairs. The most important takes place at Sonepur, on the north bank of the Ganges, opposite Patna. This was the scene of the legendary battle between an elephant and a crocodile, which Kipling adapted to explain how the elephant got its trunk. The god, Krishna, came to the elephant's aid and slew the crocodile. The great fair is held every year at the November full moon. Only 50 years ago, over a thousand elephants were exhibited at the fair. Now, with capture in India virtually banned, the numbers have declined to little more than a hundred. Hundreds of thousands of Hindu pilgrims trek to Sonepur to bathe in the sacred waters at full-moon time. They throng the dusty tracks among lines of elephants, which are gaily decorated with painted designs and embroidered covers. Each morning during the 10-day fair, the sun rises over scores of elephants bathing and being washed by their mahouts.

Not all the elephants at Sonepur are for sale. Some are exhibited by their owners as prestige symbols. They include 'killers', who have disposed of several mahouts. At the fair they are tranquillized. Ivory dealers circulate, collecting the tips of tusks, which are

BELOW:

Since the dawn of time elephants have been worshipped in India.

ABOVE: A massive tusker said to have killed 16 men on display at the Sonepur fair in northern India. ***LEFT:*** An elephant calmly allows the tips of its tusks to be sawn off for sale to an ivory dealer.

often sawn off by an elephant's owners to prevent them from causing injuries. Surprisingly, elephants stand calmly while the operation is carried out.

Would-be buyers carefully examine the feet and teeth of elephants that interest them. Hard-eyed elephant dealers negotiate, using finger codes to fix a price confidentially while hiding their hands from outsiders under a cloth. A good mature cow elephant will fetch US$10,000. But demand for elephants is dwindling. No longer do families need an elephant for transport on the Ganges flood plain. Raised roads have been built, and machines are taking over the elephant's forest jobs. But conservation agencies buy riding elephants to thrill tourists. Elephants carry the bride and groom at some weddings, and for that purpose it is still possible to rent an elephant.

Elephant dealers are like horse traders. They carefully examine the teeth of a possible purchase (above), and fix the price by exchanging finger codes under a cloth for secrecy (right).

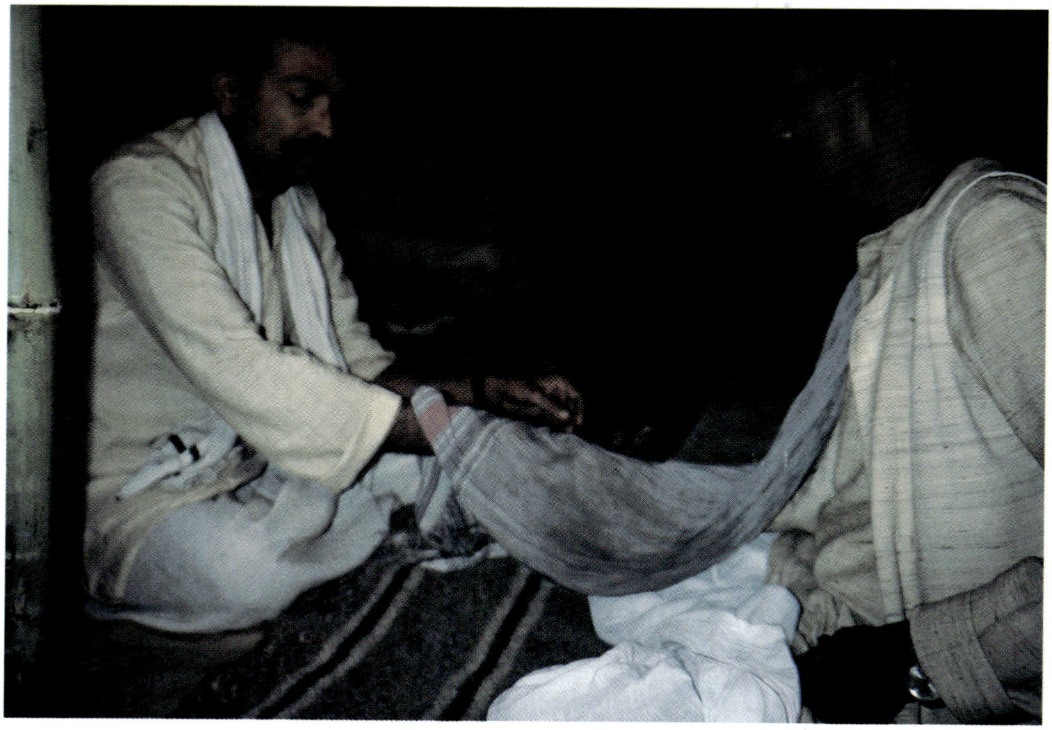

SIX

WHITE GOLD

LEFT:
Intricate carving of an ivory ship displays the delicate (some may say deadly) skill of Chinese craftsmen in Hong Kong.

ABOVE: American whalers whiled away the hours by incising designs on sperm whale teeth. The carvings are called 'scrimshaws'.

Prehistoric artefacts show that Stone Age people already appreciated ivory. It is easy to imagine them round a camp fire whittling interesting forms from the tusks of a mammoth killed for food. They became highly skilled carvers, and their products represent the earliest known sculpted work of humans. Female figurines in ivory from 25,000 years ago have been found across Europe and northern Asia, from France to the heart of Siberia. Scholars named the squat but intriguing figures 'venuses', and suggested that they represented a mother goddess of fertility. One tusk fragment from St Germain in France has a mammoth carved on it.

Huge amounts of mammoth ivory still exist. Basset Digby, a Fellow of the Royal Geographical Society, visited the New Siberian Islands before the Russian Revolution. He reported that, every spring, the melt waters eroded cliffs and river banks and caused landslips, revealing tusks. He saw a hunter's hoard consisting of 'mammoth tusks by the dozen, by the score – hundreds and hundreds of them, cairn upon cairn, stack upon stack. Tons and tons of prehistoric ivory.' Among them he found a monster tusk of 3.9 m (12.8 ft), and several nearly as long. The Yakut people used mammoth ivory for such mundane things as pipes, fishhooks, harpoon and arrow heads, in addition to crude figurines and caskets. Considerable quantities of mammoth ivory are still marketed by the USSR.

Carvers in Egypt were already turning elephant tusks into images of lions and baboons before the time of the pharaohs. With elephants easily available in Sudan and the Horn of Africa, as well as in Syria, ivory was widely used during the 3,000 years of the pharaoh dynasties. Among the many examples of Egyptian ivory work is a 2,900-year-old ivory statuette of Cheops, whose tomb is the Great Pyramid, near Cairo. The remains of other ancient civilizations, such as the Assyrian, Etruscan, Hittite, Minoan, Mycenaean and Phoenician, include much evidence of the use of ivory.

Elephant ivory has been called 'white gold.' Its beauty lies in its texture. Essentially it is just a prolonged tooth composed of dentine, not very different from those of humans. However, only the tip of a tusk has enamel, which

LEFT:
A rare tusker is the pride of its owner in Sri Lanka.

soon wears off. The tusk has a tree-like bark, which the carver removes. Elephant dentine is compact and laid in successive layers throughout life. Exceedingly fine tubes spiral through the tusk, giving it high elasticity. In minute spaces between them there is organic matter, which makes the ivory easy to work and to polish. Ivory from different areas varies in hardness, translucency, ability to withstand temperature changes without cracking, and tendency to turn yellow. The ivory of African bush elephants is soft compared with that of the forest elephant, and tends to have lateral lines. Sperm whales, narwhals, walruses and hippopotamuses also have 'ivory' teeth, but they lack the special qualities that make elephant ivory so esteemed.

Only gold, also easily worked, has equalled ivory in human esteem. They have frequently been combined in works of art. The Ancient Greeks covered monumental statues of Zeus and other gods at Delphi with ivory plates, which were then ornamented with gold and gold leaf.

The Bible tells us that Solomon had a magnificent ivory throne covered with gold. According to the First Book of Kings: 'The top of the throne was round behind, and there were stays on either side on the place of the seat, and two lions stood beside the stays. Twelve lions stood there on one side, and on the other, upon the six steps.' The chronicler was undoubtedly right when he declared: 'There was not the like made in any kingdom.'

The biblical account also recalls that Solomon's navy regularly brought him ivory, along with gold, silver, apes and peacocks, probably from India. In the *Song of Solomon,* the sensuous aspect of ivory, which is ideal for representing the human skin in statues, appears in the words: 'My beloved . . . his belly is as bright as ivory covered with sapphires', and 'Oh, prince's daugher . . . thy neck is as a tower of ivory.'

The Romans made extensive, even profligate, use of ivory. Thrones were made of ivory; senators sat on ivory benches; there were ivory beds, chariots and carriages, book covers and bird cages, even floors. Caligula's horse was honoured with an ivory stable and manger. Seneca had 500 tables with ivory legs.

Ivory has been favoured from early times for religious objects. Christian crucifixes, rosaries, figurines, altar pieces, tablets and panels, bas-reliefs, relics and incense caskets, are matched among Hindus and Buddhists by arrays of deities and representations of epic stories. Non-religious works include depictions of tales of chivalry, love and hunting; dagger hilts; powder horns; chesspieces; garnishings for toilet articles such as hairbrushes and mirrors; and inlays on furniture and firearms. Ivory was chosen for billiard balls, handles of knives and coffee pots, dice and piano keys. Expressions such as 'rattle the ivories' for dicing, and 'tickle the ivories' for piano playing came into currency. Although plastics have replaced many uses of ivory, concert pianists still like to play on ivory keys, because they are slightly absorbent and thus facilitate grip.

In China, ivory has been highly valued and delicately carved for several thousand years. Ivory combs, hairpins

FAR LEFT:
Chinese craftsmen are specialists in carving sets of ivory balls, one inside the other, from a single block.

and wine vessels have been found at Angshang (Hunan Province) in remains of the Yin-Shang dynasty of 3,000 years ago. There was no shortage of supply at that time because there were still Asian elephants in a large part of the country south of the Yellow River, and in adjoining Southeast Asia. Tribes under Chinese suzerainty paid tribute in ivory; Chinese traders brought ivory from India; and mammoth ivory was imported from Siberia.

ABOVE: A single file of elephants on a tusk makes a popular souvenir. ***LEFT:*** Cutting the tusks ready for carving is the work of experts, who ensure that little is wasted.

Arab traders, who had contacts with the elephant-rich lands of India and Africa, came into prominence in the ivory trade with China in the first millenium AD. They settled in the ports, especially Guangdong (Canton). In the eighth century AD, the Arab traveller, Al Mas'udi, reported that Chinese officials rode in ivory-veneered palanquins. Chinese ivory work became highly popular in Europe when the country was opened up to Western traders after the mid-nineteenth-century opium wars. It included delicate, concentric ivory balls, carved one inside the other; statuettes of emperors and empresses; inlays on lacquer boxes and furniture; intricate palaces, ships and scenic displays (not to mention elephants themselves, individually and marching in line) – and chopsticks.

The art of ivory carving reached its peak in China. Handed down from father to son, it survives today, despite the migration of many skilled people to Hong Kong during the upheavals which have convulsed the country during this century. Guangdong craftsmen recently carved 41 concentric balls depicting a classical story of the lady Chang-o who flew to the moon in search of immortality, and returned to Earth with her fairy maidens.

In Japan, in the eighth century AD, officials had to carry ivory emblems, which appear to have been made in China. Other objects of the period found in Nara, the first capital of Japan, also appear to be of Chinese origin. They include red-and-black-painted measures, sword scabbards and game pieces. In due course, Japanese craftsmen themselves raised ivory carving to

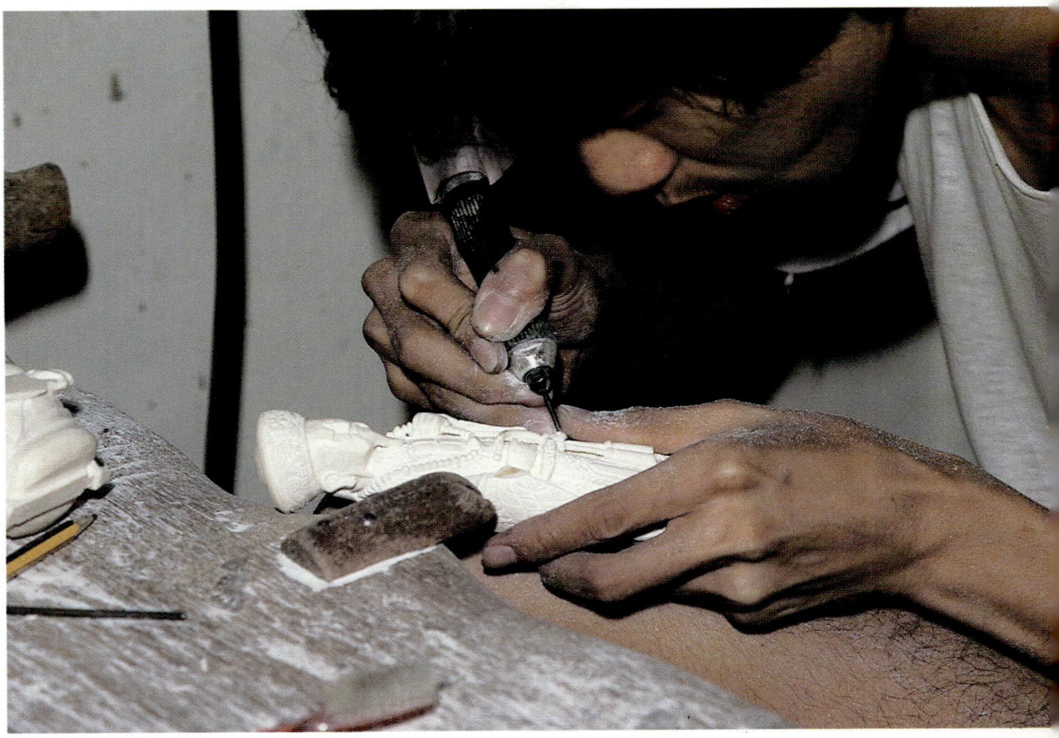

ABOVE: Today's ivory craftsmen make use of modern tools, such as dentist's drills.
LEFT: However, feet can provide a useful vice.

great heights. Particularly notable are objects known as 'netsuke.' These are small ivory carvings of extremely varied subjects, which formed toggles on kimonos or were fitted to a sash to hold swords, fans, medicines boxes and pipe cases.

Present-day Japan is the world's largest user of worked ivory. More than half is carved into personal seals or *inkan*, which represent a person's or a company's signature on documents, such as bank cheques and legal contracts. Among several types of seal used for different purposes, the most important is a person's *jitsuin*, which is registered with the government as an official mark on coming of age. It is estimated that at least 70 million ivory *inkan* are in existence, and that 2 million are produced every year.

ABOVE AND RIGHT: Rich pavilions and elaborate ships show all the mastery of Chinese ivory craftsmen.

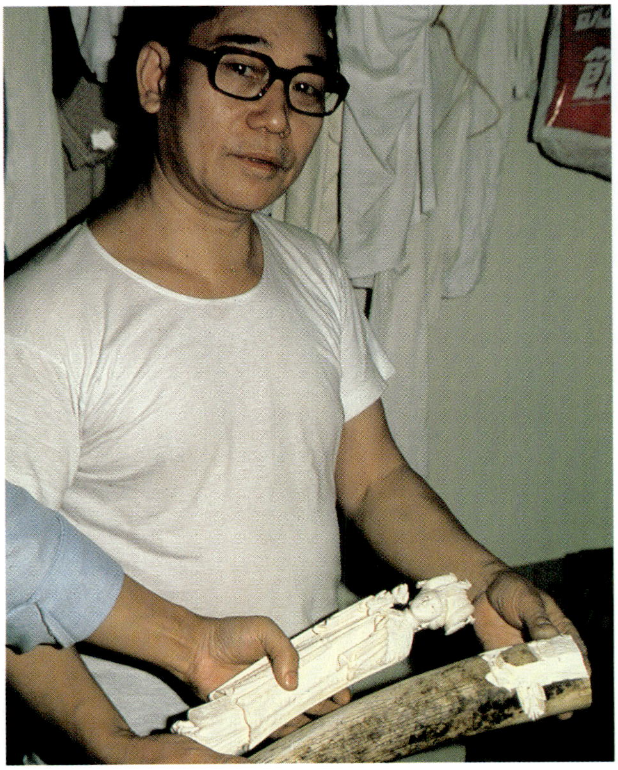

TOP, LEFT AND ABOVE:
Almost any theme can be turned into an ivory work of art. Carvers usually specialize, some in kings and queens, others in ships or pavilions or animals.

Ivory carving is an ancient art in India, going back 4,000 years to the Indus Valley civilization. Apart from the use of ivory for religious images and the common uses elsewhere of ivory, Indian craftsmen became experts at inlays in rosewood, sandalwood, horn and tortoiseshell. Miniature painting on ivory developed in the eighteenth century, and is attributed to European influence.

In sub-Saharan Africa, people lived among millions of elephants. Even without the ivory taken from elephants hunted for meat, there was plenty available from those that died naturally. People wore ivory bracelets, earrings and necklaces. Tusks were even used in quantities to make fences.

While traditional ivory carvers in Africa and Asia continue to produce works of art, demand for souvenirs by millions of tourists has promoted crude work. Shops around the world are now filled with knick-knacks of little or no artistic value. This sad development comes at a time when the ivory trade and the greed of speculators are encouraging massive poaching, which threatens the very existence of elephants.

ABOVE: Tusks surrounded by carved ivory in Hong Kong.
ABOVE RIGHT: Small pieces of ivory are turned into bangles, brooches and earrings, here on sale at appropriately-named Yvoire in France.

SEVEN

BLOODY IVORY

LEFT:

Flyers over Africa look down on the scattered, whitened bones of elephants poached for their ivory.

'Large elephant teeth are now rarely found, except in India, the demands of luxury having exhausted all those in our part of the world.' So wrote the Roman, Pliny the Elder, in AD 77. He was referring to the Mediterranean basin, where elephants in north Africa were on the verge of extinction, and had already gone from Syria. Pliny did not know of the millions of elephants roaming sub-Saharan Africa.

Several factors led to the decline of the Mediterranean's elephants. A climate change brought arid conditions, which reduced the vegetation and made the area unsuitable for elephants. Habitat was also destroyed as forests were cut to build ships and houses. But an important reason was the insatiable appetite for ivory of the Mediterranean peoples, especially the Romans. Just as happens today, ivory hunters sought the biggest tusks. As the venerable tuskers disappeared, so the hunters turned to younger elephants with smaller tusks. The Romans also captured elephants in large numbers for circuses. By about AD 200, elephants had disappeared from the Mediterranean region.

It was probably not the first time that humans had been largely responsible for wiping out an elephant species. Prehistoric hunters have been implicated in the disappearance of the mammoths and mastodonts in Eurasia and the Americas only 10,000 to 12,000 years ago. South American peoples were probably involved in wiping out Cuvier's elephant during the early years of the Christian era.

When Portuguese, British and other European traders began to explore the

LEFT:
Scottish explorer David Livingstone declared that control of the ivory trade would end the associated 19th-century slave trade.

coasts of Africa in the sixteenth century, ivory was one of the commodities they sought. The slave trade to provide labour for plantations and mines in Latin America, the West Indies and the southern states of the USA quickly became inextricably mixed with the ivory trade. Africans were rounded up and used to transport ivory from the interior to the coast, where both were sold. Records show that big tusks fetched a better price than the slaves who had carried them. The slave trade concentrated at first on West Africa, whence men, women and children were transported across the Atlantic. A British naval blockade of the slave ships in the early nineteenth century led to the collapse of the west African ivory trade. But ivory continued to flow to Europe in smaller amounts from Portuguese-controlled Angola to the south.

On the east coast, the Portuguese established a flourishing ivory trade with India and China, where they had settlements at Goa and Macao. Through the Portuguese connection, Indian traders started to settle in Africa and became in due course the principal handlers of ivory. Tusks were brought to the east coast from central Africa by slaves, who were shipped not only to America but to Mauritius and Reunion islands, where the French were establishing sugarcane plantations. The Americans were deeply involved in ivory as well as slaves. The USA was the first country to establish a consulate in Zanzibar, in 1836, because of trade prospects.

As the crusade against the slave trade developed, the ivory trade was identified as the key to its destruction. Missionary explorer David Livingstone declared in 1863: 'Get possession of the

BELOW:
Scattered bones of poached elephants on an African plain.

RIGHT: A Zambian guard stands by the carcass of a poached elephant, whose tusks have been ripped out. *BELOW:* Vultures devour the flesh of poached elephants in the Virunga National Park in Zaire.

ivory trade . . . and you render the trade in slaves unprofitable.' General Charles Gordon, appointed by the khedive of Egypt as governor of Equatoria (southern Sudan), struck a blow at the heart of the slave trade by prohibiting commerce in ivory.

In southern Africa, immigrant Dutch farmers and French Huguenots found a wildlife paradise, with abundant elephant herds. Hunting ivory as they pressed northwards to establish settlements, they had almost wiped out elephants south of the Zambezi by 1900. Ox carts carried the tusks to the ports.

By the end of the nineteenth century, the European powers, represented at first by pioneering traders and explorers, had established their vast African dominions. The slave trade had ended, but the ivory trade flourished. London, the main trade centre for most of the century, was overtaken by Antwerp when King Leopold established the Congo Free State and harvested the ivory from its teeming elephant population. When World War I broke out in 1914, the ivory trade was at its height, with 1,000 tonnes leaving Africa annually. This could have involved the deaths of 50,000 elephants a year. The four-year conflict virtually destroyed the trade, which was not to return to comparable levels for some 60 years. Ivory was still in demand for knife handles, billiard balls and similar mundane uses, but, in general, it fell from public esteem, like many pre-1914 fashions.

Ivory slowly returned to favour after World War II. In 1950 recorded world imports totalled 28.72 tonnes. Three years later, they had risen tenfold. Im-

LEFT:
The leader of an anti-poaching squad examines elephant remains.

ABOVE:
Remains of poached elephants in the Luangwa valley, Zambia.

ports continued to increase over the next 20 years to reach 485 tonnes in 1964, 537 tonnes in 1967, and 631 tonnes in 1970. In 1973, the US government abandoned the gold standard for the dollar. Gold prices, which had been rising since before World War I, shot up. The price of ivory per kilogram nearly doubled. Ivory imports in that year rocketed to 1,236 tonnes. People were turning to traditional havens for their fortunes in the face of financial instability and worldwide inflation.

In Africa, Indians, who had long controlled the ivory trade, were forced to leave newly independent countries. Many invested their personal fortunes in ivory in order to transfer them elsewhere. Africans took over the trade. With demand soaring and high rewards offered, poachers launched a massive offensive on Africa's elephants. 'White gold' was now truly to become 'bloody ivory'. First to suffer were easily available herds on the open plains of East Africa. Tracks cleared for tourists in national parks, and growing road transport networks, made access to concentrations of elephants and transport of tusks easy for the poachers. The days of traditional African elephant hunters using poison-tipped spears and arrows were over. Poachers now were armed with modern rifles, and, as time went by, with Soviet-style AK-47 automatics, which were flooding in for new armies and a host of rebel movements. As the big tuskers fell, guns were turned indiscriminately on whole herds of females and their babies. From the air, the savanna could be seen littered with bleached bones.

LEFT: Ivory seized from poachers is kept in a government store in Lusaka, Zambia.
BELOW: This tusker in the forest in Manas National Park, in north-east India, faces a new threat from tribal invaders.

Uganda, once a show-piece of African nature conservation, fell under the sway of Idi Amin. Despite his proclamations of interest in his country's natural heritage, he took no steps to halt a genocidal assault on the abundant elephant herds. Surveys by members of the authoritative African Elephant Specialist Group of the International Union for Conservation of Nature and Natural Resources (IUCN) revealed that, of 15,000 elephants in Uganda's Kabalega Falls National Park in 1966, only 1,200 survived in 1980. This was a measure of the slaughter throughout the country.

In neighbouring Kenya, poaching and a severe drought wiped out more than half its elephant population between 1970 and 1977. There were still some 65,000 elephants in 1981, but by 1987, the number had dropped to 21,000.

The poaching wave swept southwards through Africa. Northern Tanzania followed Kenya and Uganda. It still had over 200,000 elephants in 1981. Since then poachers have killed half of them. In the Lake Manyara National Park, where Iain Douglas-Hamilton unravelled the social organization of elephants in the 1960s, only four females aged over 30 remained in 1989, compared with one or more older females in three-quarters of the herds in 1980. As elsewhere, the older elephants with the largest tusks were gunned down.

Further south, the poachers attacked Zambia's plentiful herds. In the six years between 1981 and 1987 the elephant population was reduced from 160,000 to 43,000.

Zaire and Congo are the strongholds of the forest elephant. The dense habitat has made population census near impossible, and most estimates are little more than intelligent guesses. Nevertheless, the flow of distinctively hard ivory from the forest elephant reaching international markets has provided evidence of heavy poaching.

ABOVE:
Captive tuskers may be hard to handle, but are valued for their great strength.

The former Portuguese colonies of Angola and Mozambique have been wracked by bitter civil wars. Both had large elephant populations. Elephant specialists estimate that poachers have reduced Mozambique's elephants from 54,800 in 1981 to 18,600 in 1987. No estimates have been made for Angola. However, Angolan ivory has been found passing through South Africa. The UNITA rebel movement (Union for the Total Independence of Angola) has been accused of selling ivory to finance its arms purchases and operations. UNITA spokespeople have denied this.

Asian elephants have suffered less than their African cousins from ivory

poachers because many bulls and all females are without tusks. But they have not been immune; most of the big tuskers in southern India have gone. Hong Kong authorities have seized tusks shipped from Burma, where long-standing tribal rebellions provide incentives to use ivory to finance operations. However, one of the rebel movements, the Karen National Union, has accused the military government of killing elephants, citing an announcement in 1989 of the sale of ivory worth $25,000 to foreign firms. In Thailand, thieves have sawn off the tusks of tame elephants.

It was not until 1989 that the Western world and Japan finally shed their indifference to the massacre being enacted in Africa and Asia. Only then did the governments propose a complete ban on the ivory trade. A new chapter in the story of elephants and ivory was opened.

ABOVE:
Ivory poaching is not a threat to the Indian elephant as a species because only males have tusks, and not even all of them. But tuskers do get poached – as this one in southern India.

EIGHT

ELEPHANT CRISIS

LEFT:
Elephants may seem big on the ground, but from the air they can be hard to spot.

ABOVE:
Elephants have gouged the massive trunk of this baobab tree, which may be over 1,000 years old.

Fear that elephants could be hunted to extinction has only become acute in the past 20 years. A decline in the ivory trade from 1914 until after World War II allowed elephant populations in Africa to recover from the effects of decades of hunting. In Zimbabwe, elephants were well protected. Their numbers increased tenfold in 70 years. By the 1960s, the number of elephants in some parts of Africa even appeared excessive. In Kenya's 20,700-km^2 (8,000-sq. mile) Tsavo National Park, 20,000 elephants were destroying the trees on which they and other wildlife depended. Acute droughts exacerbated the damage. Many animals, including 330 black rhinoceroses, died of starvation. Fears arose that the elephants were turning Tsavo into a desert. Three hundred were shot for examination by scientists studying the Tsavo problem. They recommended that 2,700 more be eliminated to lessen pressure on the vegetation. In the event, poaching solved any problem of overpopulation of elephants in Tsavo. It was one of the first areas hit in what was to become a continent-wide offensive against the elephants for the sake of their ivory.

Elephant specialists were handicapped in warning governments of their concern, because no one knew how many elephants there were in Africa, and even exactly where they were to be found. Poaching was clearly reducing numbers in some places, but in others there appeared to be more elephants than ever before. Without details of numbers in various regions, it was impossible to assess population trends.

It may seem strange that the numbers of such a large animal were not known. But elephants are spread over a vast area, where a physical count is impossible. In open country, such as savanna and light woodland, aerial counts of elephants in selected areas can be made. Densitities in various types of habitat are then calculated, and the results extrapolated for larger areas. Even then, many elephants may be missed and allowance has to be made for a wide margin of error. Moreover, at least a third of Africa's elephants live in dense equatorial forests, where they are seldom seen. Scientists are still striving to find methods of counting them to improve on intelligent guesses.

By the mid-1970s, alarm became more widespread as ivory flooded out of Africa and the price rocketed. IUCN, with support from WWF, commissioned a pan-African elephant census by its newly established African Elephant Specialist Group. The chairman, Iain Douglas-Hamilton, who had been among the first to raise the alarm, criss-crossed the African plains in his light aircraft, counting elephants where they were visible. Simultaneously, he questioned governments and wildlife specialists to obtain local information. In 1979, he estimated a total of 1.3 million elephants in Africa, far more than anyone had thought. Nevertheless, Douglas-Hamilton declared that numbers were declining and that the African elephant was threatened by the ivory trade.

Not everyone was willing to accept this conclusion. Ivory specialist Ian Parker prepared a detailed report on

LEFT:
When elephants are crowded into a small area they can destroy their own food source by stripping the bark and thus killing trees.

the history of the trade and its recent impact on Africa's elephants for the US Fish and Wildlife Service. He insisted that the trade was not a threat, and that the ivory offtake was sustainable, even though elephant numbers were declining in some areas. He pointed out that Africa's human population was rising fast, and spreading into elephant habitat. This, he declared, accounted for any decline. The apparent increase in protected areas arose because herds took refuge there.

Despite Douglas-Hamilton's census, it was clear that precise information was still lacking for three-quarters of the estimated 1.3 million elephants. He suggested that the vast expanse of equatorial forest in the Zaire basin probably contained about 500,000 elephants. Other experts said there might even be 3 million. New studies and surveys were commissioned. Douglas-Hamilton complained bitterly that the great tragedy was being enacted in Africa while complacency reigned.

Meanwhile, ivory continued to pour out of Africa during the 1980s, despite a drop in value. In 1983 exports again matched the 1,000 tonnes of 1973. The full significance of this for elephants became clear from measurements of the average tusk weight in the major market of Hong Kong. From 9.65 kg (21.28 lb) in 1978 it declined to 5.39 kg (11.9 lb) in 1982. From this it could be calculated that the 1,000 tonnes exported in 1973 probably represented the death of about 55,000 elephants. But nearly 100,000 must have died to produce the same gross weight in 1983. Poachers were increasingly slaughtering females and immature elephants with small tusks. Observers flying over the African plains sometimes saw more carcasses than live elephants.

As concern grew, the secretariat of the Convention on International Trade in Endangered Species of Wild Fauna and Flora, known as CITES, now commissioned a new study of the ivory trade by Rowan Martin of Zimbabwe's Department of National Parks and Wild Life Management. He put the total number of elephants in 1986 at about 1 million, and calculated that Africa's elephants could sustain an annual harvest of 750 tonnes of ivory a year *if* they were efficiently managed. In the existing situation, when very few elephant populations were managed at all, he estimated the sustainable harvest at only 329 tonnes a year. Yet in that same year nearly 800 tonnes of ivory reached world markets.

Fortunes could be made from ivory at every level, from the poacher in the field, through chains of middlemen, to the big importers in centres like Hong Kong, Singapore and Japan. Because the trade was officially controlled by permit under CITES, shipments of illegal ivory accompanied by apparently legal documents could double the value. Smuggling, forgery, bribery and corruption spread throughout Africa and wherever ivory moved. It reached the highest levels of government. The most notorious country was Burundi, a tiny state in central Africa, whose only elephant was in a zoo. Burundi is ideally placed, close to the borders of the countries with Africa's largest elephant populations. It became a channel for illegal ivory. Early attempts by CITES to block the Burundi trade failed because Japan continued to allow imports, mostly via Belgium.

Another major ivory outlet was Sudan. When the government there imposed a ban in 1983, traders moved in force to Burundi. Dubai (one of the United Arab Emirates), Singapore and the Portuguese colony of Macao, near Hong Kong, became staging posts on the Japan route.

In response to international pressure, Japan finally imposed strict import legislation in 1986. In the same year, CITES established a system by which each country could set its own annual quota for ivory exports. Countries that

BELOW:
It remains a mystery why elephants sometimes ravage one patch of forest without harming an adjoining, similar area. Fire has also swept through this Zambian forest, compounding the elephant damage.

ABOVE: Trees reduced to skeletons by elephants in Hwange National Park, Zimbabwe.

did not announce quotas were banned from trade. The CITES secretariat had to confirm the authenticity of ivory shipments before an importing country could receive them. In this way legal trade could be monitored. Essentially, governments subscribing to CITES would only allow import of ivory from approved African producers. Traders left with huge illegal stockpiles were permitted to register them by the end of 1986 so that they could be legally marketed. The aim was to mop up illegal ivory, and clear the way for a well-managed legal trade. Under pressure from CITES, Burundi agreed to abide by the convention and to register the tusks it held – over 17,000. Every newly registered tusk tripled or quadrupled in value. A CITES consultant involved said later that he was offered bribes of $750,000 to allow more tusks to be included. He and his family were threatened. But the registration was completed. Despite this, a few weeks later, 30 tonnes of illegal ivory reached Singapore through Burundi. The United Arab Emirates refused to succumb to pressure to halt illegal ivory traffic and withdrew from CITES. Other governments were warned not to deal with Dubai. Nigeria, which had no ivory export quota, became a major trans-shipment point for raw and carved ivory because of its lax import and export controls.

Zaire has the largest elephant population in Africa. During the 1980s, it may have provided one-third of total exports from the continent, despite a ban on trade in raw ivory in 1982 and on elephant hunting in 1984. Its only legal ivory comes from seizures, and recorded exports between 1986 and 1989 were only 80 tonnes. But it is a thickly forested country, two-thirds the size of the USA, with corruption at all levels. It is impossible to prevent vast quantities of ivory passing the frontiers to neighbouring countries. A local carving industry also uses around 70 tonnes of ivory every year, and its products are smuggled out.

All over the world contraband ivory was discovered. Belgian customs officers found 10 tonnes in two containers labelled 'beeswax'. With a value of one million dollars, it was in trans-shipment to Dubai. In Zambia, 6 tonnes were uncovered in a secret compartment of a truck on the way to Burundi. A shipment of malachite in Lisbon concealed 1.5 tonnes. An ambassador leaving Tanzania was caught with 184 raw tusks, 24 partly worked tusks and 82 carved figures, as well as ivory necklaces and walking sticks.

However, improved controls, rising prices and campaigns against buying ivory goods in western Europe and the

BELOW: Elephants sometimes try their strength against trees and fell them, for no apparent reason.

ABOVE: Counting large animals, such as elephants, from the air may sound easy. But spotting them in the bush can be difficult. An elephant is just visible near the centre of the river bank.

USA were having some effect. Statistics showed a marked drop in imports of African ivory from 1986. Growing public pressure led to bans on imports of raw ivory by the European Community, the USA, Canada, Switzerland, Australia and Taiwan. Japan imposed a temporary ban. The USA, importer of one-third of the worked ivory in trade, used the threat of an import ban to force Hong Kong, its main supplier, to extend its prohibition on imports of raw ivory to semi-worked ivory. Hitherto, traders were legitimizing illegal raw ivory by having it partly carved by Chinese craftsmen in Dubai before shipping it to Hong Kong.

Meanwhile, the slaughter of elephants continued. The ivory trade was like water, always finding its way through an incomplete dam, and the poachers in the field were still able to get good returns. The price of ivory again soared in 1986 and 1987, reaching a record $123 per kg ($56 per lb).

Tanzania and Kenya now took the lead. Alarmed by their rapidly diminishing elephant populations, they proposed a ban on all international commerce in ivory. The World Wide Fund for Nature gave its full support. The president of Kenya, Daniel Arap Moi, dramatized the situation by publicly setting fire to 12 tonnes of tusks, worth $2.5 million on the international market.

Strong opposition came from countries in southern Africa with thriving elephant populations. Zimbabwe keeps growing numbers in check by culling to a level consistent with the vegetation available. Ivory and hide are legally exported in return for hard currencies. Local people get elephant meat. This encourages them to accept the presence of elephants, which can seriously damage crops. South Africa's elephants are almost all in the strictly controlled Kruger National Park. Culling controls numbers, the ivory is legally marketed, and meat is processed in modern abattoirs. Botswana too has flourishing herds. For these countries an end to the ivory trade means considerable economic loss.

When government delegations to CITES gathered in Lausanne, Switzerland, in October 1989 to tackle wildlife trade problems, the Ivory Trade Review Group told them that Africa's elephant population had been more than halved in 10 years and now numbered 609,000. The massacre, first of bull elephants, and then of females, was disrupting the elephants' social structure and breeding, and thereby threatening the potential for recovery. The death of orphaned calves accounted for one in three of all elephant deaths. Moreover, the elimination of elephants was affecting the biological wealth of Africa's savannas and forests because elephants help disperse seeds and, by their feeding, create a patchwork of habitats which benefit both wildlife and people.

By now public opinion was fully aroused. Hundreds of children paraded outside the Lausanne conference hall with banners pleading with the delegates to save elephants by stopping the ivory trade. Arguments raged for a full week on the floor of the conference and in late-night meetings. Zimbabwe led the opposition to a ban on the grounds that countries which had looked after their elephants were being penalized because of the failure of the majority. Finally, 76 nations voted for a ban, with 11 against and 4 abstentions.

Technically, this meant that the African elephant was moved from Appendix II of CITES, which permits controlled international commerce in wildlife products, to Appendix I, which prohibits all such trade.

The interests of the opposing countries were not overlooked, for the CITES resolution specified that the next conference, in 1991, would consider whether elephant products in

TOP: Villages in Zimbabwe. Growing human populations need living space. In countries with elephants, the wild habitat becomes fragmented and destroyed by settlement and agriculture.

ABOVE: Agriculture in southern India.

some countries could be re-opened to trade. This would depend on their numbers, the effectiveness of conservation measures, and the degree of control of the movement of ivory through the international market. Nevertheless, Botswana, Burundi, Malawi, Mozambique and Zimbabwe announced that they would not accept the ban and would continue to trade in ivory.

While the world's attention focused on the plight of the African elephant, the status of its Asian cousin continued to worsen. It now survives in 13 countries and numbers not more than 50,000 in the wild – less than one-tenth of the 1989 African population. About half the elephants are in India. The threat of extinction hanging over the Asian elephant was recognized from the very inception of the years of negotiations which culminated in CITES in 1975. It was placed on Appendix I so that international trade in the Asian elephant, its ivory and all other products was banned. As in Africa, tuskers have been poached, but, since only a portion of Asian male elephants have tusks, this has not posed a direct threat to the species. The problem lies in the attrition of its habitat, which has been continuing for 2,000 years. In this century, human population pressure has fragmented elephant habitat almost everywhere by converting forests to agriculture and settlement, and disrupting traditional migration patterns. Herds now find fields of rice and sugarcane where they remember forests. Inevitably, they try to feed on such delicacies. Violent clashes arise when people try to drive them away. People and elephants are injured and even killed.

As the world enters the last decade of the second millenium, many people wonder whether the elephant, 'nature's great masterpiece' in the words of the poet John Donne, will survive in the twenty-first century.

ABOVE: Scientists and wildlife managers use computers to work out the most effective conservation measures for elephants.

BELOW: Agriculture encroaches on a Sri Lankan elephant reserve.

NINE

CONSERVING ELEPHANTS

LEFT:
An Asian elephant mother with her calf.

Elephants are wide-ranging animals, needing vast living areas. In the distant past in sub-Saharan Africa, when people were comparatively few, they had ample room. Africans killed elephants for meat, and made use of the ivory, but until the advent of European traders and colonialists, the herds were not threatened. Greed for ivory then led to an orgy of hunting, which virtually wiped out elephants south of the Zambezi and throughout much of West Africa. Herds elsewhere were severely depleted.

By 1900, most of Africa had fallen under colonial sway. Reaction against the wildlife massacre led to the first protection laws. Nature reserves were established in Natal and Transvaal in South Africa in 1897 and 1898. In 1925, the Albert (now Virunga) National Park in the Belgian Congo became the continent's first national park. A year later South Africa upgraded the Sabie Game Reserve to the Kruger National Park. Other parks were established in the Belgian Congo, Central African Republic, Rwanda, Sudan and South Africa. Wildlife was still widespread and abundant throughout the continent, and the 1930s became the age of grand hunting safaris by wealthy Europeans and Americans.

However, consciousness of the need to conserve Africa's wildlife led, in 1933, to an international conference in London. It concluded with a convention on the 'Preservation of Fauna and Flora in their Natural State', which was the first of its kind in the world. The convention provided for protection of rare and endangered animals.

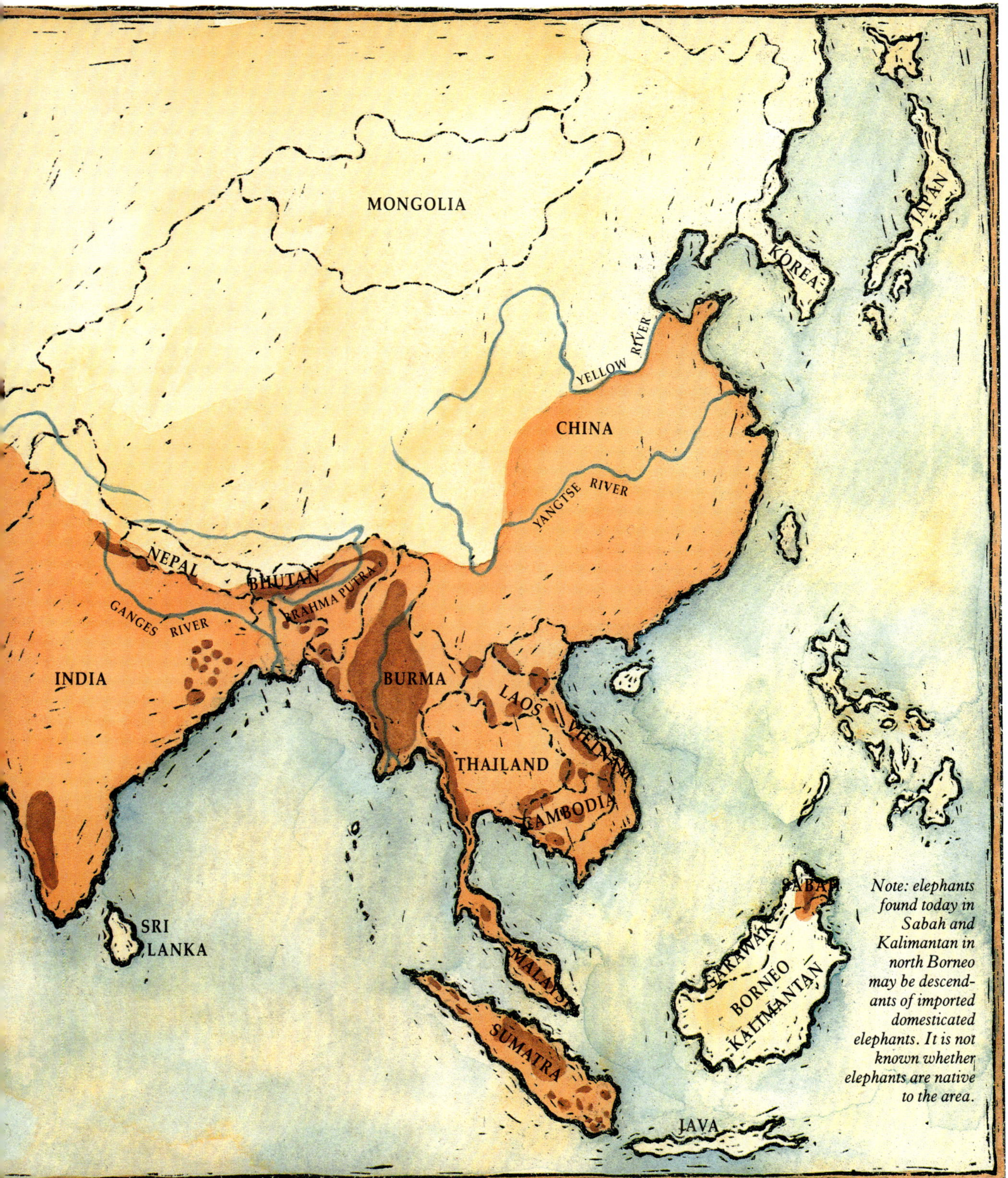

Note: elephants found today in Sabah and Kalimantan in north Borneo may be descendants of imported domesticated elephants. It is not known whether elephants are native to the area.

TOP: Secure, at least for the time being, an elephant herd rests in the midday heat.
ABOVE: Settlements fragment elephant range in Zimbabwe.

Elephants were not included as they were not considered to be endangered.

Independence came to African countries in the 1960s. The new African governments quickly showed that they too cared for their heritage. At Arusha, in Tanganyika (now Tanzania), their manifesto declared: 'These wild creatures, amid the wild places they inhabit, are not only important as a source of wonder and inspiration, but are an integral part of our natural resources and of our future livelihood and well-being.' The manifesto was followed in 1968 by a convention providing for more conservation areas. Elephants were so common and widespread that no special reserves were considered necessary.

In the 1970s, when ivory poaching was reaching a threatening level, only 175,000 or 13.5% of the estimated 1.3 million elephants were in reserves. These reserves covered only 350,000 km^2 (135,000 sq.miles) or 5% of the 7 million km^2 (2.7 million sq.miles) of elephant range. Few of the reserves had adequate staff or equipment to manage and protect their wildlife, especially against gangs of well-armed, mobile poachers. African nations lacked the financial resources to improve the situation, and international aid was totally insufficient to help them.

Africa's human population has been growing at 2.5–3% a year. Elephant habitat has been reduced by 20% in the past decade, and the demand continues for more land for settlement and agriculture. It can only come from wildlife habitat. Conservation of elephants, and all wildlife, has to be reconciled with human needs. To meet the challenge, an African Elephant Conservation Coordinating Group (AECCG) was established in July 1988 to produce a plan of action. It brought together IUCN, WWF, two leading American organizations, the African Wildlife Foundation and Wildlife Conservation International, and the European Community. They decided to target 41 selected elephant populations, totalling about 250,000, for emergency protection in order to preserve the genetic and behavioural diversity of the species. At the same time, the group is exploring ways to protect the remaining 350,000, while recognizing that numbers are bound to decline in the face of human pressure.

For the African elephant, the key issue is the ivory trade. As long as there are elephants there will be ivory, a durable product that has been highly valued by humankind for thousands of years. There are still sufficient elephants to provide a sustainable harvest and meet all reasonable needs if the trade can be controlled and the herds protected. Southern African countries believe that they provide an example of effective elephant management for others to follow.

It is easy to be sentimental about elephants. The sight of whole herds, mothers and babies, being mowed down in the name of management is repugnant to everyone, including those carrying it out. But it is equally distressing to see elephants dying of starvation in ruined habitats, or being persecuted when they clash with human interests. A balance has to be struck to enable elephants to survive in the limited areas now available to them. Culling, however distasteful, is one method that provides economic benefits to local people and encourages them to accept the continued presence of elephants.

Zimbabwe's pride is its elephant management programme. Strict protection has permitted the population to recover from a mere 5,000 in 1900 to about 57,000 today. But because Zimbabwe's human population has

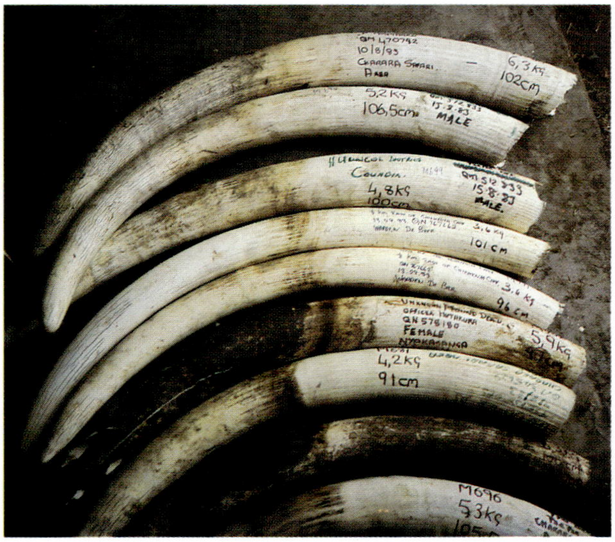

ABOVE AND LEFT: Zimbabwe's elephants are thriving under strict protection. Their numbers are controlled by culling. Tusks are all registered and sold for public benefit.

RIGHT: Hides from culled elephants in Zimbabwe are dried in the sun before tanning. The hides are turned into shoes, belts and bags.
BELOW: Almost every part of a culled elephant is put to profitable use.

also risen 10 times to reach 10 million, the elephant population has to be kept in check by culling. Sale of ivory and hide earns the country substantial foreign exchange, part of which is channelled back to the local communities where the elephants have been harvested. These communities, which also benefit from the meat, use the funds to buy agricultural equipment and to reinforce their local infrastructure. The programme is seen as essential to Zimbabwe's economy, which also earns revenue from tourists who come to see elephants and other wild animals.

Culling teams select an isolated herd and skilled marksmen shoot all of them in less than one minute. This technique prevents the operation spreading fear and panic among other elephants. The carcasses are stripped of tusks and hide, and the meat is taken quickly to processing plants. Bones, too, are collected and reduced to fertilizer. The bulk of the hide is exported after tanning, but there is a growing local industry making bags, shoes and belts for markets at home and abroad.

In South Africa, elephants in the Kruger National Park, which numbered only 10 in 1905, increased to a level which threatened large-scale vegetation changes. This in turn would affect not only the elephants, but also other animals and their habitats. The authorities decided that the population should be held at 7,500. To implement this, scientists set a cull quota of several hundred a year. Family groups consisting of females and young were darted from the air with an overdose of tranquillizer. Ground teams shot the prostrate elephants in the brain. The carcasses were then gutted and removed to an abattoir for processing. Large

LEFT AND BELOW: Deprived of their forests, elephants in Asia often become pests, raiding crops and attacking villages. They even break through brick walls searching for grain and liquor stores.

tusks were sold at public auction, and small ones by tender. Staff got two-thirds of the meat. The rest was sold to shops and restaurants in the park. Skins, feet, tails and tail hairs went to tanneries and curio manufacturers. Most of the carcass fat was used for toilet soap. Large tuskers were not shot, because of their value as tourist attractions and baby elephants in a targeted family group were usually captured for sale.

South Africa and Zimbabwe provide examples of elephant management where the species is not endangered. But elsewhere in sub-Saharan Africa, governments are confronted with massive ivory poaching, which has depleted and disrupted their elephant herds. Until this is halted and their elephant populations given time to recover, they are unable to introduce productive management schemes.

ABOVE: Villagers in north Bengal, India, stand before their house, wrecked by elephants.

Hence their call for a comprehensive ban on the ivory trade.

The Asian elephant poses different problems. International commerce in its products has been banned since 1975. They have legal protection throughout most of their range. But, confined to the most populous area of the world, the Asian elephant has to live alongside people. It cannot be entirely isolated in reserves. Coexistence often leads to confrontation.

At Naxalbari, in northern Bengal, a migrating herd of about 60 elephants regularly returns to a former forested area, which is now farmland. In a typical incident at sunset one day, the elephants line up on the edge of the remaining forest, intent on entering the crops. Facing them are hundreds of people shouting, beating tincans, and throwing stones and flares to drive them away. The matriarch leader of the herd charges. A shot in her forehead turns her away. But the diversion allows other elephants to break through the people's defences. They spend the night feasting. The dawn light spreads over devastated rice and pulse fields. A few days later the herd threatens another village.

Elephants ravage banana plantations and trample kitchen gardens. To escape them, women, children and infirm people move at night to a safe haven. The situation becomes particularly dangerous if elephants find jars of country liquor and get drunk. It seems to be an experience they enjoy, for they feed on intoxicating wild fruits.

In the course of clashes with people, elephants are often injured by muzzle-loader guns or burned by flaming missiles. This can turn them into vicious rogues, which have to be shot. Angry villagers place crude bombs, acids and poisons inside baits of banana leaves and rice, gravely injuring an elephant's trunk, the most sensitive part of its body. If it does not succumb at once, it suffers a lingering death from starvation.

BELOW: A family party about to take refuge from marauding elephants on a nearby hill.

In Sri Lanka, agricultural development has confined elephant herds in some places to small pockets of forest. There is insufficient food and water to sustain them, and so they invade the fields and village reservoirs at night. People are terrorized. Both in Sri Lanka and India some so-called 'pocketed' elephants have been tranquillized and trucked to reserves.

Because reserves large enough to isolate the 500 or more elephants that make up a genetically-viable population can no longer be maintained, IUCN's Asian Elephant Specialist Group recommends that a network of reserves be linked by forest corridors. These need be only a kilometre (0.62 mile) wide to permit the seasonal migration of elephant herds. This would ensure a large inter-breeding population and minimize conflict with people. But development authorities do not always understand the biological realities of the situation and refuse to leave corridors. Local people then suffer from elephant depredation. In other cases, people desperate for land encroach on the corridors.

To reduce clashes between elephants and humans, several measures are in use. Solid fences have limited use because they are expensive to build, and therefore cover only short distances. Deep, wide ditches are better. But, unless they are well maintained, the sides crumble and the elephants find a way across. Most effective is electric fencing designed to administer a sharp, non-lethal shock to any elephant that touches it. It has been used successfully to protect oil palm plantations in Malaysia, where elephants had been causing damage estimated at $500,000 a year. It is relatively cheap and easy to erect and move. However, electric fencing cannot be used everywhere.

Natural barriers can be created by buffer zones around reserves where

ABOVE: Villagers build fires to scare away elephants during the night.

RIGHT: Barriers can be built to keep out elephants, but only in limited areas. A spiked platform (below) guards a gate in a protective wall.

crops unpalatable to elephants are grown and no water is available. Steep-sided canals can also serve as permanent 'elephant ditches'. If necessary, crossing points can be made for elephants by building ramps to the water. Elephant tunnels under canals have not been successful.

In Sri Lanka and Sumatra, where large-scale agricultural development is under way, whole elephant herds have been moved to sanctuaries. The operation resembles the khedda method of capturing elephants, except that the herds are driven into reserves instead of stockades. Their natural response is to try to return to their old haunts.

Electric fencing or other barriers hold them until the forest has been removed and the area is no longer attractive. This method is only possible when there is sufficient reserved habitat not already occupied by elephants.

Another solution is a 'managed elephant range'. This is a forest area, adjoining a reserve, where human activities compatible with the presence of elephants are allowed. Logging, for example, can result in more food for elephants by promoting the growth of secondary vegetation. Similarly, shifting cultivation, which is common in tropical forests, can benefit elephants, provided sufficient time is left between use of a plot for vegetation to regenerate. Livestock grazing and subsistence hunting are also possible if kept to sustainable levels. In a managed range, priority has to be given to elephant conservation.

Where elephants become serious pests, and cannot be translocated,

LEFT: Deep, wide trenches can keep elephants at bay, but they deteriorate, especially in rain, if not constantly repaired.
BELOW: Elephants can be deterred by a salutary shock from an electric fence.

ABOVE: In Sri Lanka, nearly 900 elephants are being displaced by jungle clearance for human settlement.
RIGHT: The author (left) explores India's Taldapara Reserve on elephant back.

there is no alternative to shooting them. Usually, the offenders are lone bulls. Females and young can be captured, but the scope for domestication and accommodation in zoos is very limited.

Elephant populations which have been fragmented into small groups are threatened by deleterious inbreeding. Scientists therefore propose to maintain the genetic pool by translocating elephants. It is sufficient to introduce a single, unrelated bull elephant in each four-year generation.

Fortunately, large elephant populations still exist in some parts of Asia. There are over 2,000 in southern India, and equal numbers in northeastern India on the north side of the Brahmaputra River, and along the Indo-China Annamite Ridge between Cambodia, Laos and Vietnam. Provided their habitat can be maintained, their future is secure. In India, south of the Brahmaputra, a fourth population of over 2,000 is seriously threatened by destruction of habitat, which is likely to break it up into groups of fewer than 500.

Several parts of Burma are thought to have elephant populations of between 1,000 and 2,000. Their future depends on conserving the forests. In Sri Lanka, over 1,000 elephants in the southeast of the island require the maintenance of corridors between reserves for long-term survival. There is little hope of preventing the fragmentation of the large population in Sumatra's Riau Province, where people from Java are being settled. Indonesia's elephant managers hope to establish three reserves for them.

LEFT: Only on elephant back can tourists get close to rhinoceroses in Kaziranga National Park in north-eastern India.
BELOW: A baby elephant separated from its herd is bottle-fed at an orphanage in Sri Lanka.

In addition to wild elephants, Asia still has about 14,000 in captivity, principally in Burma (5,400) and Thailand (5,000). Laos has 1,300 and India 1,000. They are used for timber work in the forests, transport in difficult terrain, and as riding animals in national parks. Several hundred are kept by Hindu temples in southern India, and there are still some private owners. This large captive population is a valuable safeguard against extinction of the Asian elephant in the wild.

LEFT: Another elephant is the centrepiece at the annual festival in the south Indian village of Vengasseri.

ABOVE: A line-up of elephants graces the ceremonial dedication of an oil refinery in Cochin, south India.

TEN

IS THERE A FUTURE FOR ELEPHANTS?

LEFT:
The 'red' elephants of Tsavo in Kenya get their colour from dusting themselves with the red soil.

ABOVE:
The American bison, numbering some 70 million in the early 19th century, was nearly wiped out in 50 years by uncontrolled massacre. The same could happen to elephants.

Elephants and people have several things in common. We are social creatures. We care deeply for our children, and help our fellows in trouble. Our lifespans run to around 70 years. We even suffer from some of the same diseases. But while elephants ignore people who do not get in their way, people hunt and kill elephants. People have created the elephant problem. It is up to them to solve it.

At present, ivory dominates the situation. The lust for ivory is devastating Africa's elephants. Unless it can be brought under control, elephants could become extinct in the coming century. It may seem incredible that several hundred thousand elephants could vanish so quickly. But we have only to recall the bison of the North American plains. There were possibly 70 million in the early nineteenth century. In the space of 50 years, wanton slaughter reduced them to a few hundred. In Africa, ivory poaching has already killed 500,000 elephants in the past 10 years, cutting their number by half.

Yet ivory could be the elephants' salvation if the profits from it went to local governments and people prepared to manage the herds for a sustainable harvest. At present the profits go to poachers, traders, corrupt officials and profiteers, whose interest is in making money quickly, without caring about the future. Despite the present ban on international trade, and despite campaigns against the use of ivory, it is unlikely that this beautiful substance will ever be devalued. History has shown that it vies with gold in human esteem. But while gold is mined and could theoretically become exhausted, ivory is constantly being produced. An elephant's tusks grow exponentially throughout its life, so that the biggest ivory comes from the natural death of old elephants. Ivory is long-lasting, unlike skins and other animal products, which decay quickly after death. The ideal management strategy would be to allow elephants to die naturally and then collect their tusks. This is not feasible; but harvesting of ivory and other elephant products by official culling of an approved quota in each country offers a solution. The present situation shows that if a constant weight of ivory is taken, smaller and smaller tusks are removed and more and more elephants must die to produce the same weight of ivory. This is what is happening with the present systematic poaching.

Harvesting only tusks above a certain minimum size could be successful, if the minimum allowed females enough reproductive years to maintain a corresponding increase in numbers. Elephant specialists have estimated that a compromise, whereby limits would be set both on the number of tusks harvested and on their weight, could approach the yield predicted for natural mortality. At present, such a solution is impossible in most countries. Massive poaching has disrupted the social organization of elephants. They need time to recover before a sustainable, managed yield of ivory could be envisaged.

A study by the London Environmental Economics Centre at University College, London, estimates that African nations, except Zimbabwe, have been receiving only 10–20% of the Hong Kong value of their ivory. The centre favours, in principle, a Zimbabwe proposal for the establishment of a Southern African Centre for Ivory Marketing. It recommends that ex-

BELOW:
The view from Kilaguni Lodge in Kenya's Tsavo National Park provides visitors with a timeless vision of Africa.

clusive purchasing agreements should be agreed between the primary consumer states, and that membership of what would, in effect, be an Ivory Exchange, similar to other commodity exchanges, should be based on demonstrated compliance with sound elephant management.

Apart from ivory, there is the value of elephants as a tourist attraction, which brings in revenue. Visitors to East Africa, and to Zambia and Zimbabwe, are drawn especially by the sight of the giants of the Earth in their natural habitat. In Asia, visitors enjoy riding on elephants in the jungle and watching them work and take part in colourful ceremonies.

While the African elephant could 'pay its way' in terms of ivory, the overall problem, affecting also the Asian elephant, is the clash between elephants and people for living space. Here, elephants have to give way. The demand for agricultural land and forest exploitation is such that, in future, elephants may have to be limited to reserves and to elephant ranges where compatible human actitivies are permitted. If the peoples of Africa and Asia want elephants to survive, they will have to guarantee them this living space and accept some restriction on their own activities. The decision rests with the governments of those countries that have elephants. They are all faced with serious economic problems of poverty and the need for development. Happily, the interests of people and elephants can coincide. Catchment areas for reservoirs, built for irrigation and power generation, and forests where sustainable timber extraction is carried out, are often good elephant habitat.

If people outside Africa and Asia also want elephants to survive – and it appears that most do – then they have a duty to provide the necessary assistance. The plight of the African elephant has spurred the European Community to commit a large sum for a conservation programme. Various governmental aid organizations are also showing interest, and the general public has already donated large amounts to charitable organizations. It is still not sufficient. And little provision has so far been made for the Asian elephant.

The measured tread of a herd of elephants on the move against a backdrop of forest and mountain conveys a sense of timelessness. This is the world as it has been for millions of years, from long before *Homo sapiens* appeared. Several hundred species of elephants evolved in that time, only to disappear. Our ancestors contributed to the extinction of some of them. Today, just two species remain. Their fate is in our hands.

ABOVE:
Wild elephants are a major tourist attraction in some parts of Africa, earning valuable foreign exchange for their home countries.

AFRICAN ELEPHANT POPULATION 1989		
	ELEPHANTS	ELEPHANT RANGE KM^2
CENTRAL AFRICA		
Cameroon	22,000	253,000
C.A.R.	23,000	346,000
Chad	2,100	202,000
Congo	42,000	214,000
Equatorial Guinea	500	23,000
Gabon	74,000	249,000
Zaire	112,000	1,420,000
CENTRAL AFRICA TOTAL	***276,000***	***2,707,000***
EAST AFRICA		
Ethiopia	8,000	139,000
Kenya	16,000	413,000
Rwanda	50	3,000
Somalia	2,000	56,000
Sudan	22,000	372,000
Tanzania	61,000	501,000
Uganda	1,600	16,000
EAST AFRICA TOTAL	***111,000***	***1,500,000***
SOUTHERN AFRICA		
Angola	18,000	458,000
Botswana	68,000	93,000
Malawi	2,800	19,000
Mozambique	17,000	246,000
Namibia	5,700	141,000
South Africa	7,800	23,000
Zambia	32,000	211,000
Zimbabwe	52,000	114,000
SOUTHERN AFRICA TOTAL	***203,000***	***1,305,000***
WEST AFRICA		
Benin	2,100	20,000
Burkina Fasa	4,500	36,000
Ghana	2,800	22,000
Guinea	560	11,000
Guinea Bissau	40	400
Ivory Coast	3,600	51,000
Liberia	1,300	17,000
Mali	840	50,000
Mauritania	100	6,000
Niger	440	6,000
Nigeria	1,300	29,000
Senegal	140	10,000
Sierra Leone	380	3,000
Togo	80	7,000
WEST AFRICA TOTAL	***18,000***	***268,000***
TOTAL FOR AFRICA	***608,000***	***5,780,000***

(Based on the report of the Ivory Trade Review Group. Totals rounded to the nearest 1,000)

SELECT BIBLIOGRAPHY

Cumming, D. H. M. and Peter Jackson, 1982. *The Status and Conservation of Africa's Elephants and Rhinos.* IUCN, Gland (Switz).

Cumming, D. H. M., du Toit R.F. and S. N. Stuart, 1989. *African Elephants and Rhinos: Status Survey and Conservation Action Plan.* IUCN, Gland (Switz).

Douglas-Hamilton, Iain and Oria, 1975. *Among the Elephants.* Collins, London.

Eltringham, S. K., 1982. *Elephants.* Blandford Books, Great Britain.

Hanks, John, 1979. *The Struggle for Survival: The Elephant Problem.* C. Struik, Cape Town.

Hasthi. *Newsletter of the IUCN Asian Elephant Specialist Group,* IUCN, Gland (Switz).

Ivory Trade Review Group, 1989. *The Ivory Trade and the Future of the African Elephant.* Ivory Trade Review Group, International Development Centre, Oxford.

Jackson, Peter, 1982. *Elephants and Rhinos in Africa: A Time for Decision.* IUCN, Gland (Switz).

Lahiri-Choudhury, D. K., 1988. 'The Indian Elephant in a Changing World'. In *Contemporary Indian Tradition.* Smithsonian Institution Press, Washington.

Martin, Esmond Bradley, 1985. *The Japanese Ivory Industry.* WWF Japan, Tokyo.

Pachyderm, *Newsletter of the IUCN African Elephant and Rhino Specialist Group.* IUCN, Gland (Switz).

Parker, Ian and M. Amin, 1983. *Ivory Crisis.* Chatto and Windus, London.

Payne, Katherine, 1989. 'Elephant Talk'. In *National Geographic* (August 1989). National Geographic Society, Washington DC.

Santiapillai, Charles and Peter Jackson, 1990. *The Asian Elephant Action Plan.* IUCN, Gland (Switz).

Scullard, H. H., 1974. *The Elephant in the Greek and Roman World.* Thames and Hudson, London.

Seidensticker, John, 1984. *Managing Elephant Depredation in Agricultural and Forestry Projects.* The World Bank, Washington DC.

Sikes, S. K., 1971. *The Natural History of the African Elephant.* Weidenfeld and Nicolson, London.

Sukumar, R., 1989. *The Asian Elephant: Ecology and Management.* Cambridge University Press, Cambridge.

Toynbee, J. M. C., 1973. *Animals in Roman Life and Art.* Thames and Hudson, Great Britain.

Traffic Bulletin. Wildlife Trade Monitoring Centre, Cambridge.

INDEX

Note: page numbers in *italics* refer to illustrations